Battlefield Weapons Systems
& Technology, Volume IX

MILITARY DATA PROCESSING AND MICROCOMPUTERS

Other Titles in the Battlefield Weapons Systems and Technology Series

General Editor: Colonel R G Lee OBE, Military Director of Studies at the Royal Military College of Science, Shrivenham, UK

This new series of course manuals is written by senior lecturing staff at RMCS, Shrivenham, one of the world's foremost institutions for military science and its application. It provides a clear and concise survey of the complex systems spectrum of modern ground warfare for officers-in-training and volunteer reserves throughout the English-speaking world.

Volume I	Vehicles and Bridging—I F B Tytler *et al*
Volume II	Guns, Mortars and Rockets—J W Ryan
Volume III	Ammunition, including Grenades and Mines—K J W Goad and D H J Halsey
Volume IV	Nuclear, Biological and Chemical Warfare—L W McNaught and K P Clark
Volume V	Small Arms and Cannons—C J Marchant Smith and P R Haslam
Volume VI	Command, Communication and Control Systems and Electronic Warfare—A M Willcox, M G Slade and P A Ramsdale
Volume VII	Surveillance and Target Acquisition Systems—W Roper *et al*
Volume VIII	Guided Weapons, including Light Unguided Anti-tank Weapons — R G Lee *et al*
Volume IX	Military Data-processing and Microcomputers—J W D Ward and G N Turner
Volume X	Basic Military Ballistics—C L Farrar and D W Leeming
Volume XI	Military Helicopters—P G Harrison

For full details of these and future titles in the series, please contact your local Brassey's/Pergamon office

Related Titles of Interest

GARNELL	Guided Weapons Control Systems (2nd ed.)
HEMSLEY	Soviet Troop Control. The Role of Automation in Military Command
LEE	Introduction to Battlefield Weapons Systems and Technology
MORRIS	Introduction to Communication, Command and Control Systems

MILITARY DATA PROCESSING

AND MICROCOMPUTERS

J. W. D. WARD

and

G. N. TURNER

Royal Military College of Science, Shrivenham, UK

BRASSEY'S PUBLISHERS LIMITED
a member of the Pergamon Group

OXFORD · NEW YORK · TORONTO
SYDNEY · PARIS · FRANKFURT

U.K.	BRASSEY'S PUBLISHERS LTD a member of the Pergamon Group Headington Hill Hall, Oxford OX3 0BW, England
U.S.A.	Pergamon Press Inc., Maxwell House, Fairview Park, Elmsford, New York 10523, U.S.A.
CANADA	Pergamon Press Canada Ltd., Suite 104, 150 Consumers Rd., Willowdale, Ontario M2J 1P9, Canada
AUSTRALIA	Pergamon Press (Aust.) Pty. Ltd., P.O. Box 544, Potts Point, N.S.W. 2011, Australia
FRANCE	Pergamon Press SARL, 24 rue des Ecoles, 75240 Paris, Cedex 05, France
FEDERAL REPUBLIC OF GERMANY	Pergamon Press GmbH, 6242 Kronberg-Taunus, Hammerweg 6, Federal Republic of Germany

---

First edition 1982

Library of Congress Cataloging in Publication Data

Ward, J. W. D.
Military data processing & microcomputers.
(Battlefield weapons systems & technology;
v. 9)
Includes index.
1. Military art and science—Data processing.
2. Microcomputers. I. Turner, G. N. II. Title.
III. Series.
UG478.W37 1982 355'.0028'54 82-13159

British Library Cataloguing in Publication Data

Ward, J.W.D.
Military data processing & microcomputers.—
(Battlefield weapons systems & technology; v. 9)
1. Weapons systems—Data processing
I. Title II. Turner, G.N. III. Series
623.4'028'5404 UF500

ISBN 0-08-028338-1 (Hardcover)
ISBN 0-08-028339-X (Flexicover)

In order to make this volume available as economically and as rapidly as possible the authors' typescript has been reproduced in its original form. This method unfortunately has its typographical limitations but it is hoped that they in no way distract the reader.

The views expressed in the book are those of the authors and not necessarily those of the Ministry of Defence of the United Kingdom.

Printed in Great Britain by A. Wheaton & Co. Ltd., Exeter

Preface

The Series

This series of books is written for those who wish to improve their knowledge of military weapons and equipment. It is equally relevant to professional soldiers, those involved in developing or producing military weapons or indeed anyone interested in the art of modern warfare.

All the texts are written in a way which assumes no mathematical knowledge and no more technical depth than would be gleaned from school days. It is intended that the books should be of particular interest to army officers who are studying for promotion examinations, furthering their knowledge at specialist arms schools or attending command and staff schools.

The authors of the books are all members of the staff of the Royal Military College of Science, Shrivenham, which is comprised of a unique blend of academic and military experts. They are not only leaders in the technology of their subjects, but are aware of what the military practitioner needs to know. It is difficult to imagine any group of persons more fitted to write about the application of technology to the battlefield.

This Volume

This book attempts to introduce the fighting soldier to sufficient computer technology to allow him to understand how computers work, be aware of their battlefield applications, and to play a full part in the specification of the requirement for future systems. It is therefore intended for those who wish to widen their professional military knowledge.

Shrivenham. March 1982 Geoffrey Lee

Acknowledgements

In preparing this book, our greatest difficulty has been in attempting to present a large number of facts at the correct level for the target reader. Computer terminology and jargon are initially difficult to grasp and technical books abound; our problem has been to provide explanations in everyday terms for military officers. We have only therefore been able to take a fairly high level look across the breadth of our subject, but the bibliography offers areas for further reading which may be necessary to enable a deeper understanding; this is particularly true in Chapter 3, where we have used examples to help the reader to understand the breadth of range of high level languages and their application, but the examples are necessarily of a general nature.

Our grateful thanks are due to Mr. Adrian Marks for his help in commenting upon what we have written and to the following for so willingly providing information and photographs:

Fred Eldridge, David Almond of Software Sciences Ltd. Graham Williams, Richard Curtis, Errol Hay, Bryan Hubble of Marconi Space & Defence Systems Ltd. Brian Packman of Plessey Ltd. Richard Blake of the Royal Military College of Science. Richard Hughes of Ferranti Computer Systems Ltd. G. Smith of Racal-Comsec Ltd. David Ashman of Intel Corporation (UK) Ltd. T. F. Taylor of Cossor Electronics Ltd. Richard Mann of Texas Instruments Ltd. Peter Eveleigh of Marconi Radar Systems Ltd. Peter Walker of Print, Promotions and Publicity Ltd. for Commodore Business Machines UK Ltd.

Shrivenham. March 1982

John Ward
Graham Turner

Contents

List of Illustrations

1.
Introduction

INTRODUCTION

To most soldiers, computers are complicated electronic machines that perform a multitude of intricate calculations at inconceivable speed, and it is by no means clear how such machines can be useful in handling arithmetic or information or controlling the working of some other device. The usual explanations which are offered are often confusing in that they involve jargon and are orientated to a particular commercial product.

There is no doubt that, since the invention of the computer, the media and the computer industry have managed to highlight the more sensational capabilities and applications as they evolve. Computers can process data in order to play chess, read books, store fingerprints, speak and draw pictures. Of course these applications are awesome but the reporting of them seldom takes into account the vast amount of effort expended in their development, nor the multitude of rather unglamorous, routine tasks which the great majority of computers are required to complete. Even the transferring of a simple accounting problem to computer will involve lengthy and very detailed work; an abstract task, such as the automation of some information handling processes will be a complex, manpower intensive, expensive project.

The aim of this book is to provide an explanation of automatic data processing (ADP) for the layman and to describe how this new science is being applied on the battlefield.

HISTORY OF COMPUTING

Fingers were the first aids to calculation; the word 'digit' derives from the Latin for 'finger' and the decimal system probably came into existence as the result of man having ten readily available digits.

The first known calculating machine is the abacus. A primitive abacus consisted of a tray of sand in which lines were drawn and on which pebbles were placed to represent numbers; the Latin for 'pebble' gives us the word 'calculation'. The most familiar abacus is the Chinese version consisting of beads on wires; examples can still be found in use in the Far East.

Pascal invented the first machine to perform the four fundamental arithmetic operations of adding, subtracting, multiplying and dividing in 1642. In 1671 Leibnitz developed a multiplication mechanism which could multiply directly, rather than by repeated addition, added it to Pascal's design, and recommended it for scientific use as well as the commercial role for which Pascal had produced it. Many competitors followed, as scientists saw the capabilities and labour and time saving potential of these computers. Towards the end of the nineteenth century the development of a family of mechanical analogue computers started; now they operate electronically and are normally used in engineering.

Also during the nineteenth century the first attempts were made to build a digital computer; Charles Babbage considered that the various scientific tables of information available at that time were incomplete and inaccurate. He developed a 'Difference Engine' which could calculate the tables and print them out, thus avoiding the normal human errors. Babbage obtained government support for further development and the Difference Engine concept was used by the Registrar General to produce life tables on which life insurance estimating was based. His work progressed to the design of the 'Analytical Engine' as he realised that it was possible to build more general-purpose, more automatic machines; the operation of the automatic control was to be by a string of perforated cards, a method invented in France in the eighteenth century and used successfully on the Jacquard loom. Babbage was never able to build his Analytical Engine, because of the limitations of the engineering techniques available and perhaps the grandeur of his design, but his aims of accuracy, speed and 'economy of intelligence' were undoubtedly correct.

In an analogue computer, the problem to be solved is set up as a model in which the behaviour and performance of the component parts is imitated by electronic circuits. The problem and its model correspond very closely and measurements can be easily transferred from one to the other. Numbers are represented by the magnitude of some measureable quantity (eg continuously variable electric voltages). However, there are severe limitations in practice:-

Accuracy is poor, due to the limitations of the electronic circuits used.

Analogue computers must often be dedicated to the particular role and there are some difficulties in producing multi-role machines.

Computing units are accurate only over a limited range.

Data cannot be stored within the computer in any significant quantity.

In contrast, digital computers use individual indicators or units for each digit in a number, which allows precision to be increased by merely enlarging the machine to allow it to handle more digits; the only limit to this increase is cost.

It was during the 1939-45 war that a requirement was found to compute artillery firing tables automatically for the American Army; in 1943 a machine called the Electronic Numerical Integrator and Computer (ENIAC) was built to carry out this task. The technology to produce this machine had been available for some years but the war highlighted the requirement.

In 1944, J von Neumann set to work with the designers of ENIAC and the result was a report which outlined the design of the modern electronic computer. Most important was the concept of storing sets of instructions (programs) within the computer in numerical form thus allowing the computer to operate on a program just as on the numbers to be processed.

Fig. 1.1 The prototype 'Mark 1' computer at Manchester University

One of the earliest 'stored program' computers was built in Manchester University and started operating in June 1948. The first stored program computer for commercial use was called UNIVAC 1 and was designed for the American Bureau of Census in 1951. The first computer procured specifically for stock control was a modified version of that at Cambridge and was built by J Lyons & Co in 1953; it was known as the Lyons Electronic Office (LEO).

Since these early machines were built, computers have increased in power, speed and reliability; numerous applications have been found and are still being found. The microprocessor has meant smaller and very much cheaper products but no new principle has been introduced.

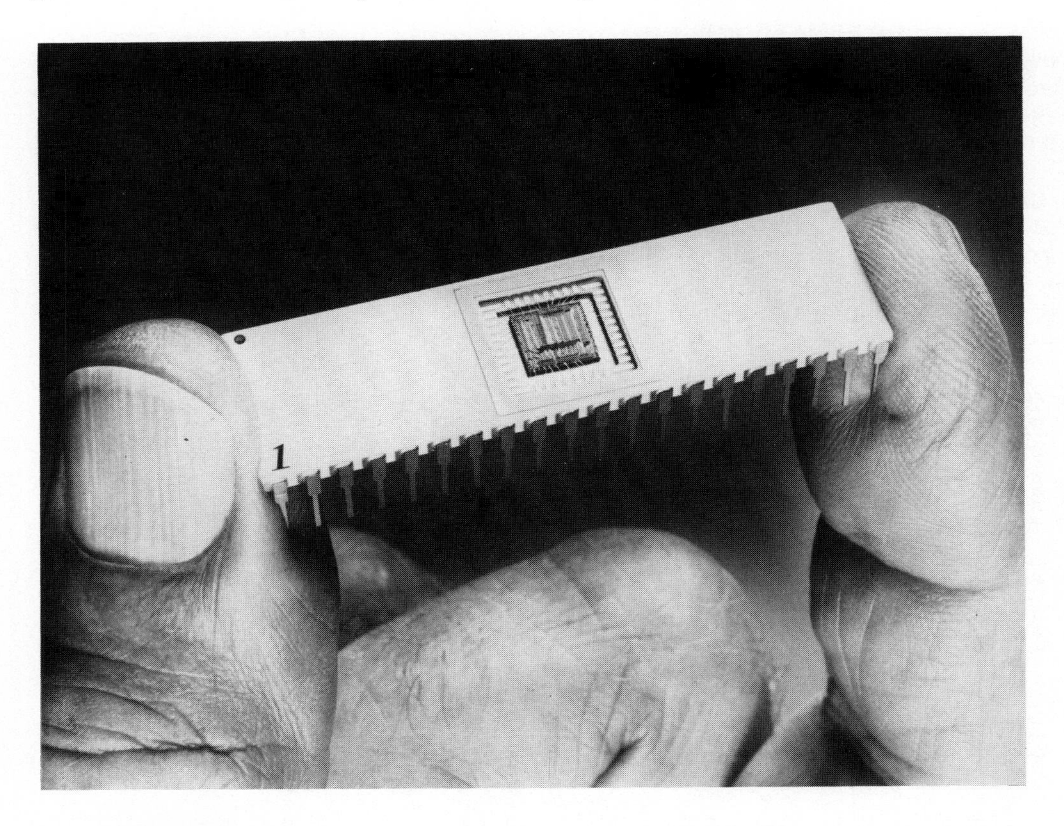

Fig. 1.2 A microprocessor

HOW A DIGITAL COMPUTER WORKS

Hardware and Software

A digital computer system can be considered in parts; the HARDWARE and the SOFTWARE. The hardware refers to the physical components which are the various electronic and magnetic devices. The software refers to the PROGRAMS which control the operation of the hardware. The choice of hardware will of course depend upon the application to which a computer system is to be put, but the flexibility will largely depend upon the software, which is at least as important as the hardware for successful operation.

The Concept of a System

It is important that a computer should be considered as a system to process information. A system can be defined as a group of objects related or interacting to form a unit. At a low level this can mean specifically a particular interrelated collection of components such as a computer system and at a higher level, for instance, a complete data processing system; the term can include all the machines, operators, managers and their overheads necessary to carry out an overall task. Thus a system can be at any level and, indeed, every system will form a part of another, larger system.

Representation of Information within the Computer

Two fundamentally different types of information exist within the digital computer whenever it is active; these are PROGRAMS and DATA. It is important to be quite clear as to the distinction between these two types of information.

A program is a set of instructions which the computer stores and executes in a prescribed sequence to carry out some task such as to update a payroll, solve a set of simultaneous equations, or compute the ballistics of a gun. The program is written by a human being and a typical example of the sort of thing a program might do is:

for every man in a regiment:

find	BASIC SALARY
calculate	INCOME TAX
compute	NET PAY
print out	NAME, RANK, and NET PAY

Before the computer can execute a program to carry out this sequence of operations every step must be defined in minute detail; for instance where do the figures for BASIC SALARY come from and what is the formula for the TAX calculation? The program must then be stored in the computer's own memory in a form that the computer understands and all necessary starting data such as BASIC SALARY must be made available.

The data for a program consists of the information on which the program is to work. Starting data, such as BASIC SALARY in this example, is normally read into the computer's memory via an input unit such as a card reader. Intermediate data, or working data, such as TAX is already within the computer. Final data such as NET PAY is usually transmitted back to the outside world via an output unit such as a printer. The only way that the computer can tell the difference between a program and data, because they are both stored as digits, is by their location in the computer memory or store; each location in a store has a unique ADDRESS.

As a general rule when information is being transmitted from place to place within the computer, or when it is undergoing processing, it is in electrical form and when it is being stored it is in electrical or magnetic form. It is convenient to represent all information, whether programs or data, within the computer in binary form, because of the physical nature of the devices used for handling the information. Binary notation is a system for representing numbers in which the 'base' or 'radix' for each digit is 2. In this system numbers are represented by the two digits '0' and '1'. In the same way that in the decimal system a displacement of one digit position to the left means the digit is multiplied by a factor of 10, so in the binary system similar displacement means multiplication by 2. Thus the binary number '110101' represents '53' in the decimal system as follows:

	x 2 = 32	x 2 = 16	x 2 = 8	x 2 = 4	x 2 = 2	= 1	32 16
Binary	1	1	0	1	0	1	4 1
Decimal Equivalent	32	16		4		1	= 53

Fig. 1.3 Binary/decimal conversion

A transistor can be in one of two states, ON or OFF, or a magnetic device can be magnetised in one direction or the other; such reliable 'two-state' devices abound. ON can be made to represent 0 in the binary notation and OFF can be made to represent 1. The arithmetic required to process information held in binary form is relatively simple. For humans, who are used to manipulating ten digits, a binary code is an awkward way of expressing information and although the computer can be programmed in its own binary MACHINE CODE by a human programmer this is a most laborious process. Normally a program is written in a language close to English, called a HIGH LEVEL language, and this is translated by the computer into the computer's own machine code using another established program called a TRANSLATOR or COMPILER.

The basic unit of information within the computer is the binary digit, a 0 or a 1, and this is referred to as a BIT. Just as humans find it convenient to have other levels of information above the basic symbol of a single digit or letter so does the computer. There are two important units apart from the bit; first the BYTE or CHARACTER which is a conveniently sized collection of bits for moving data around inside a computer system but may not be large enough to represent a complete item of information. It consists normally of 8 bits in modern computers, eg:

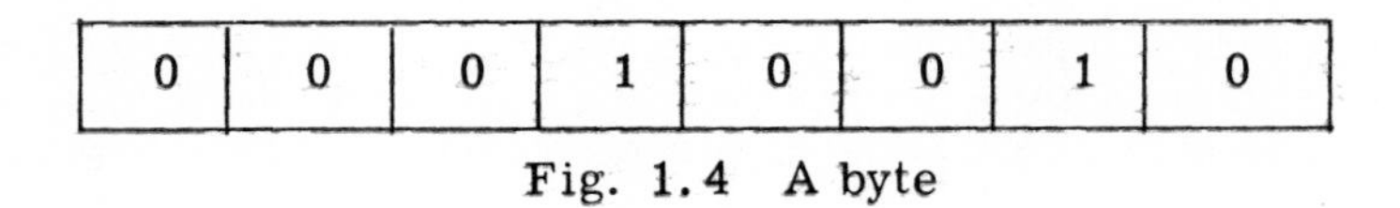

0	0	0	1	0	0	1	0

Fig. 1.4 A byte

The second is the WORD which is the normal usable unit of information. In most modern general purpose computers the word length is either 16, 24 or 32 bits (ie 2, 3, or 4 bytes).

Data within the computer usually represents numbers; for example BASIC SALARY would probably be input as a number of pounds, dollars, or cents etc. There are two main ways in which numbers can be converted into digits and stored:

FIXED POINT, where the number is represented by a single set of digits; the value of the number depends on the position of the digits with respect to a fixed, predetermined position for the decimal point (eg 50.7).

FLOATING POINT, where the number is represented by two sets of digits known as the MANTISSA or FIXED POINT PART and the EXPONENT (eg 5.07×10^1, where 5.07 is the fixed point part and 10^1 is the characteristic).

The use of floating point arithmetic allows numbers of larger ranges of magnitude to be stored more economically and thus calculations to be carried out to higher consistent accuracy.

HARDWARE

Modular Concept

Computers are more easily understood if they are broken down into modules and considered in three main areas: these are the CENTRAL PROCESSOR UNIT (CPU), BACKING STORE, and INPUT/OUTPUT devices as shown in Fig. 1.5. The backing store and input/output devices are often grouped together and are generally known as PERIPHERALS.

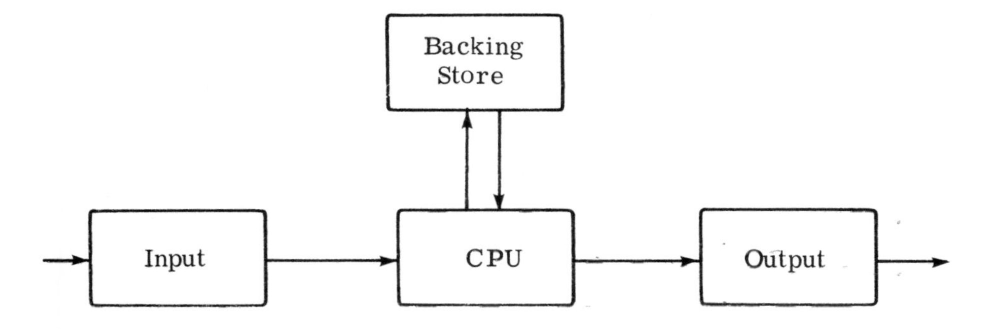

Fig. 1.5 The basic block diagram

The CPU

The CPU is the nerve centre of the digital computer, since it co-ordinates the activities of the other units and performs the arithmetic and logical operations. It can be considered in three parts (see Fig. 1.6). First the ARITHMETIC and LOGIC UNIT (ALU) which, as its name implies, performs the arithmetic and logic upon the data put into it from the main store in accordance with appropriate program instructions. It works at a very high speed and its operation is measured in microseconds. It follows that if there is to be no wastage of processing time the programs and data to be used must be available within microseconds. This availability, or immediate access, to data and programs is achieved by holding the necessary information in the adjacent main store.

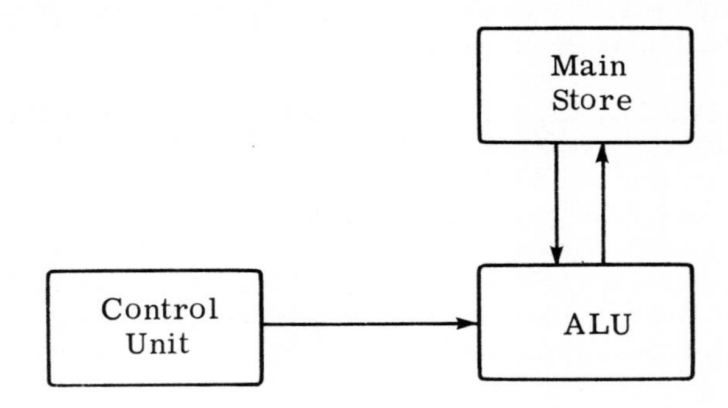

Fig. 1.6 The main parts of the CPU

The MAIN STORE is an array of devices organised to hold data and programs grouped as bytes or words in a series of locations. The devices commonly used are magnetic or electronic. Most magnetic devices used are ferrite core magnets, the direction of the magnetism in each core dictating whether the bit held in the core is a 0 or 1. The electronic devices are semi-conductor devices which act as a memory; it is built out of two-state electronic circuits. In special-to-purpose computers, such as the conventional pocket calculator, where it is essential that the programs cannot be altered once they have been set in during manufacture, a permanent store called READ ONLY MEMORY (ROM) can be employed; many larger systems also may well contain elements of ROM. Temporary data storage is normally carried out by the use of 'read/write' memory, usually referred to as RANDOM ACCESS MEMORY (RAM). It should be noted that, in many systems, main store is considered separately to the CPU.

The CONTROL UNIT is the third part and it operates upon instructions from the programs to initiate and control the appropriate computer operations, not only for the CPU but for the other hardware components of the computer system.

The Backing Store

The backing store holds data and programs which are not immediately required, in a cheaper form than using the more expensively constructed main store. Data and programs are held in the backing store in some magnetic form, but access to these programs and data, unlike that in the main store, often involves the use of mechanical devices. For this reason backing stores operate at a very much slower speed than main stores. Typical backing stores are the MAGNETIC DRUM, the MAGNETIC DISC, and MAGNETIC TAPE and they are described in more detail in Chapter 2.

The magnetic drum was the earliest device and is now rapidly going out of use; data and programs are recorded on the magnetic surface of a drum which is continuously rotated at a high speed, past fixed recording and playback heads. Thus data and programs are available in the time the drum takes to make one revolution. The magnetic drum is a fast type of backing store. In the case of the ıagnetic disc, data and programs are recorded on the magnetic surface of a num- ›er of discs which are rotated at high speed past moveable recording and playback ıeads. The access time is governed by the time the head takes to move to the ıppropriate track and the time the disc takes to move one revolution. It is clear hat it takes longer to obtain data from the magnetic disc than from the magnetic lrum. These forms of storage are described as DIRECT ACCESS stores or sometimes CYCLIC ACCESS stores.

The operation of magnetic tape differs little in principle from that of domestic ape recorders. The recording and playback head is fixed and the magnetic tape ıolding the data and programs is moved past the head to the appropriate sections of tape. Even though the tape is moved at high speeds, the access time may be very long and is measured in seconds, or even minutes. This is described as a SERIAL ACCESS store. Although the access to data on magnetic tape is very much slower than for the magnetic drum and disc, the tape is very much cheaper.

Input and Output Devices

Input and output devices are necessary before the computer can be set to work to allow programs and data to be entered into the main and backing stores. Similarly when the computer has finished its task, the result must be displayed in a form intelligible to man. A computer system therefore includes input and output devices, essentially designed to match the characteristics of man and machine. There are combined input and output units or they may be separate.

Devices which cater for both input and output are commonly known as 'interactive' devices. They often take the form of a VISUAL DISPLAY UNIT (VDU) which is similar to a television screen and is a flexible device normally based on a cathode ray tube. It may be a diagrammatic and/or an alphanumeric (ie letters and numbers) display. With the addition of a keyboard and suitable circuitry the device can be used to enter data to the computer. Often a print out of the output is required, in this case a HARD COPY TERMINAL may be used. This device, which may be a teleprinter or a teletype, can be connected to the computer to receive or transmit data in the form of electrical signals. The input speed is naturally limited by the typing speed of the operator, and the output speed is limited by the design of the device. An advantage of this terminal is that hard copy of both input and output can be provided. ACOUSTIC devices are also being introduced; they are still in their infancy and it may be many years before they are in widespread service. They convert human speech directly into electrical signals acceptable to the computer, and for output will produce a passable imitation of the human voice.

Input-only devices can take many forms. The CARD READER accepts cards once they have been punched with a combination of holes to represent letters, numerals or special symbols; converts the information represented by the pattern of holes into electrical signals suitable for input to the computer. Alternatively a PAPER TAPE READER may be used; it accepts a paper tape which again has been punched across its width with a row of holes, each row of holes representing a character. A more sophisticated process is MAGNETIC INK CHARACTER RECOGNITION (MICR) to read letters and numbers specially printed in magnetic ink. Even more advanced is OPTICAL CHARACTER RECOGNITION (OCR) which is a method of recognising printed characters.

Output-only equipment can also use punched card and punched tape. The CARD PUNCH is operated by electrical output signals from the computer to produce the appropriate combination of holes in a card. Similarly the PAPER TAPE PUNCH provides a punched tape. Another output-only device is the LINE PRINTER; they are commonly electro-mechanical, but by using electrostatic printing techniques much higher speeds can be achieved. A useful facility can be the employment of a GRAPH PLOTTER, which prints out in a pictorial or graphical form on paper or transparent material.

SOFTWARE

Definition

The speed, accuracy and reliability of computer systems is such that man is now able to undertake work previously impossible by virtue of its volume and complexity; he is also relieved of much of the lengthy and tedious work associated with modern life. However, the computer, like any other machine, is dependent upon the ingenuity of man for its design and operation. It can only perform tasks in accordance with the programs previously prepared by man, and for this reason the software is as vital as the hardware for the successful operation of a computer system.

Software is the collective term used to describe the various programs which control the operation of the hardware; these programs appear to the computer as a string of bits, which are referred to as machine code. For man to write all his programs in machine code would be not only tedious but very open to error. To overcome this problem special languages have been developed which allow programmers to write their instructions to the computer in simple words and abbreviations.

Languages

Languages are usually developed for particular applications. Examples are COBOL which is the acronym for COmmon Business Orientated Language and CORAL which is Computer Orientated Real-time Application Language. The latter has been officially adopted in UK for Defence purposes. Such languages are high level languages; the closer the language approaches natural English the higher is the level. Programs written in high level language are called SOURCE PROGRAMS and those written in machine code are called OBJECT PROGRAMS. To translate the source program into the object program the computer requires another special program called the COMPILER, which was mentioned earlier.

Multi-Programming

The flexibility and potential of a computer system depends heavily upon the ingenuity of the software design. The software can even overcome some hardware limitations, as for example in the technique called MULTI-PROGRAMMING. This technique overcomes the widely differing operating speeds of the central processor, backing store, and other peripheral devices. The CPU may work 1000 times faster than the other parts of the system, and might therefore spend much of its time waiting for the other devices to perform their tasks. Multiprogramming avoids this situation by arranging that whenever the CPU is waiting for the operation of another unit, the next task is called forward and processed by the CPU until the original program can be continued. This technique is carried out using several programs arranged in priority, switching from one program to the next every time the CPU is interrupted. Such a system is potentially complex and can often only be incorporated in computers having a large main store capacity.

Multi-Access

The speed of the central processor is so high in relation to the input/output devices, that with suitable software the central processor can share its capacity among many users. Under this system, known as MULTI-ACCESS, it appears to each user that he is having sole use of the computer. In practice the central processor is constantly switching between the programs and data from a number of users, using time-sharing techniques. There is no need for the users to be near the computers; they can be any distance away, provided that data transmission links, either radio or telephone, are available. It will be obvious that a multiaccess system implies a degree of multi-programming.

OPERATION

Concepts of Operation

The operation of a computer is normally discussed in one of two ways; the traditional method which envisages the storage devices where the data is held as the heart of the process, and then the more modern method which uses the communication between the elements of the computer as the basis for its operation. First we will look at the traditional method which is included only because readers may find it used elsewhere in older literature.

Figure 1.7 shows a conventional block diagram of a computer. It is adequate for this discussion but many of the elements may be duplicated; for example there may be a number of input devices of diverse types and many main store modules. Also architectural diagrams of individual computer systems will be much more detailed than this introductory picture. Nevertheless the principals are the same.

Traditional Block Diagram Concept

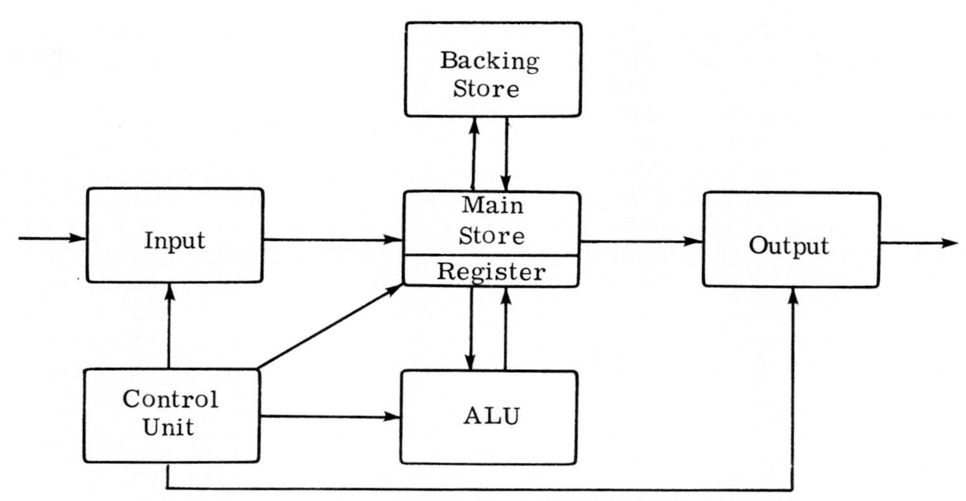

Fig. 1.7 Traditional computer block diagram

All information held by the system exists in patterns of bits, and the main store is divided into cells. Each cell holds a byte or sometimes a complete word consisting of a predetermined number of bits to be processed as an entity; for example an element of data, such as a salary or a grid reference, or a program instruction.

To describe the function of the computer it is necessary first to look at the rules which govern the interaction of the parts of the system and thus cause it to perform the required processing; these rules are the program. They must be available for use in the system and the only place where they are immediately available is the main store. As stores are designed to hold data in bit patterns then the programs must be held in the same form. There must be an indication as to what action is required and this action must be such that it can be carried out by the components when they are activated to do so. An example is the order to ADD; this information would be held as a bit pattern encoding the action. Other information that would need to be available is the whereabouts of the units of information, called the data, to be acted upon, which in this case means added. This will take the form of the addresses of store cells. Its exact nature will depend on the system architecture, or the fashion in which hardware and software are designed to interact. The complete instruction consisting of the coded operation and the addresses of the data to be called forward and worked upon will be lodged in one or more cells of the main store; this data is called the OPERAND. A description of the process to be carried out must then consist of a sequence of such instructions stored in the sequential order in which they are required to be executed. This sequence of instructions held as bit patterns is known as an object program, as mentioned earlier.

The role of the control unit is the interpretation and execution of the individual instructions of the object program by carrying them out as a sequence of tasks and initiating the appropriate operations by the other computer components. A very simplified example is as follows:

Scenario: A man sitting at a visual display unit wishes to add two numbers together with the aid of the computer. A program has already been written to solve such a problem but this particular program is not used very frequently and normally resides in the backing store; thus the man has now called it forward, using some prearranged software routine, to the main store where it is sitting waiting to start.

Sequence of events:

1. The address of the first program instruction is transferred automatically to a special store in the main store called a REGISTER.

2. The control unit fetches the first instruction from the main store using the address in the register.

3. As a result of decoding the first instruction the control unit activates the VDU to display a cue to the man such as:-

"Please input the first number to be added".

4. The control unit waits for the input of the number and then directs it to a store location which has been predetermined by the program.

5. The address of the next program instruction is transferred to the register.

6. Again the VDU is tasked to request more information such as:-

"Please input the second number to be added"

7. The control unit directs the second number to another predetermined store location.

8. The address of the next instruction is transferred to the register.

9. The control unit fetches the next instruction, decodes it, and takes the two numbers to be added from their stores and causes the ALU to add them together.

10. The next instruction now orders the result to be displayed on the VDU, printed out on a printer for reference, and to be stored in yet another store in the computer for future use.

11. Once the man is satisfied that the operation is complete he will disengage that particular program.

This is, of course, a very simplified example as instructions will, in reality, be extremely detailed; each activity may require several instructions, and the use of registers will vary from computer to computer, but it illustrates the relationships between programs, data, stores, inputs/outputs, and the control unit.

The accessing of instructions and data from the store and the performance of the execution cycle is carried out under the control of timing generators which produce clock pulses to maintain the basic timing of the electronic circuits.
The necessary interval between two accesses of the same store is known as the STORE CYCLE TIME and ranges from two or three microseconds in equipments manufactured in the 1970s to tens of nanoseconds in modern equipments. Execution time, the time taken to complete the cycle of events required to perform an instruction, will vary with different types of instruction but for simple addition would be about one microsecond in most equipments.

Bus or Highway Concept

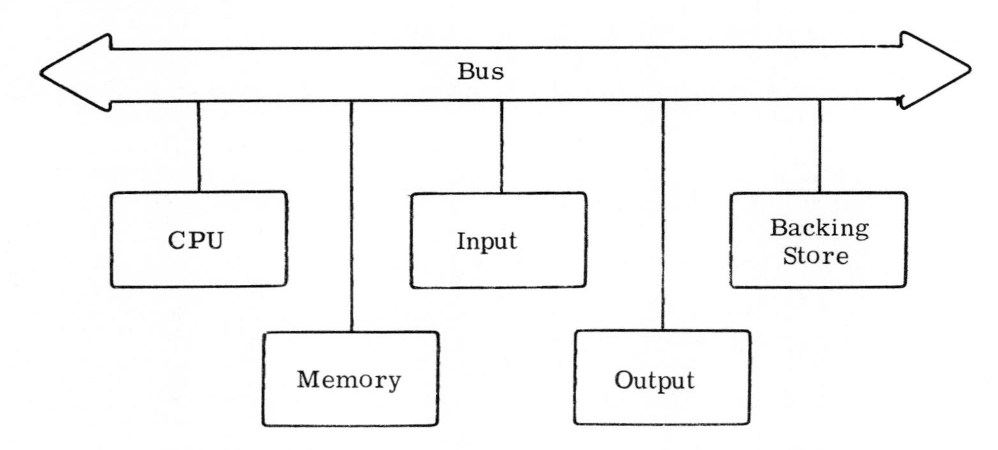

Fig. 1.8 A more modern concept of the block diagram of a computer showing how the computer might be built in practice

The use of the communication channels which connect all the components as the focus for describing the operation of the computer is relatively modern and has been reinforced by the advent of the microprocessor. Indeed the processing elements of a microcomputer are so small that the communication channels are the largest and most cumbersome components. In Fig. 1.8 the BUS or HIGHWAY is the route along which signals travel from one of several sources to one of several destinations. Attached to the bus are the same components as in Fig. 1.7 but not packaged in an exactly similar way; for instance the name 'CPU' has replaced the control unit and includes the ALU and all the control functions. The name of the main store has been replaced by MEMORY, which includes the store addressing mechanism. Input and output devices are shown as before but having similar status to backing store which is shown as just another peripheral. Each device in the figure is assumed to have its own management capability and to be able to sense when it is required to perform. As mentioned earlier this figure shows a different, modern approach to the operation of a computer and is used hereafter in this book.

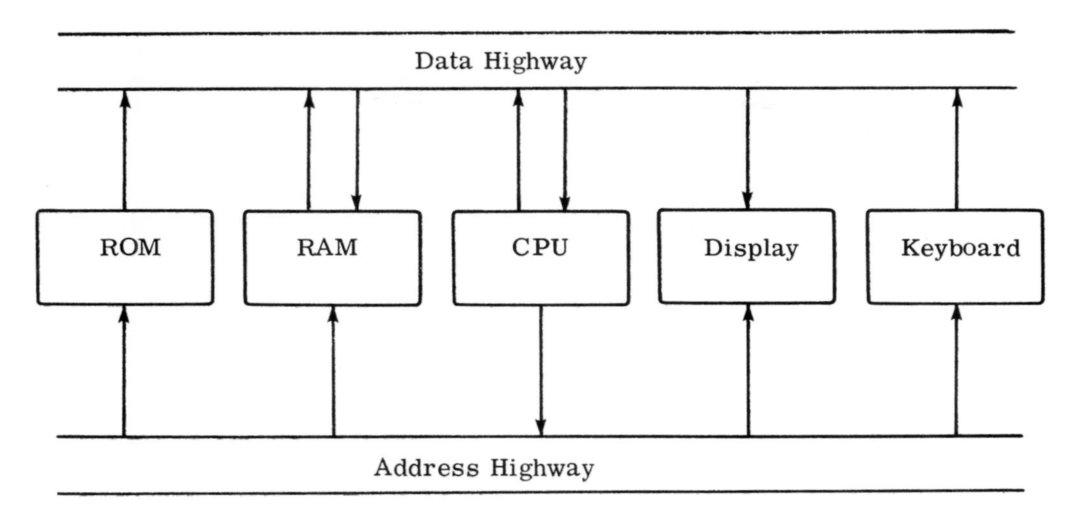

Fig. 1.9 A typical modern block diagram or schematic of a limited role computer system

Figure 1.9 shows how the schematic for a small general purpose 'desk top' calculator might be drawn. Two highways are shown, one for the CPU to address the devices and one to allow the data to flow between them. The ROM will hold the program of the set of procedures necessary to carry out all the calculations required of the machine; these are called ALGORITHMS. The RAM will be used as a working store first to hold data as it is input via the keyboard, and second to hold the partial results during the calculation processes.

PROGRESS OF COMPUTER UTILIZATION

Types of Systems

The introduction and development of the computer have outstripped the growth of its application and it is difficult to compare the potential scale of application of ADP with anything else invented by man; the nearest is possibly the art of writing itself for, although ADP cannot replace original thought it can provide widespread processing of data and present the resulting information far more efficiently than any human, in a most remarkably fast, clear, concise and accurate fashion. In order to discuss how the application of ADP has progressed, to illustrate the opportunities presented, and at the same time to give some general views on problem areas it is convenient to divide the subject into four main groups of computer application:

The more conventional data processing systems.
Scientific systems.
Systems based on communications and 'real-time' processing.
Process control systems.

Conventional Data Processing Systems

Automatic data processing systems represent a larger financial investment than the other three groups considered together. They are well established in many fields and are commonly used in commerce and administration for straightforward batch processing of large quantities of data, using conventional and well established hardware and software. Batch processing is a method of dealing with data when a number of individual items are involved; these are commonly called TRANSACTIONS, and are collected together and prepared for processing as a single unit or batch. In most cases time is not critical and there may well be a delay between the gathering and preparation of the data and the eventual processing. As the computer does not require food, sleep or relaxation time, and today very little maintenance, much of the processing can be carried out during periods when the majority of computer owners are off-duty and with minimum operator intervention. In most cases the requirement for, and function of, the system is well defined; for example payroll production, stock control, and accounting have been everyday ADP applications for many years.

Most large business organisations use computers in this role by now and the main development currently lies in the medium and small sized organisations. In the highly competitive 1980s the firm that can reduce its administrative costs to a minimum may well be the one that stays in business, and there is no doubt that computers can make a significant contribution to help. Unfortunately there are many examples of the introduction of a computer causing costs to rise; this is normally the result of a lack of knowledge and very inadequate communication. The computer industry has also been plagued by a large number of salesmen who could see a vast array of potential users before them with no knowledge of this new science but with a desperate need for assistance in handling all the information that the modern world thrives on. Consequently many users have been talked into buying equipment that does not meet their needs and in some cases has proved a positive hindrance. In a great many cases this has caused reversion to the original manual system.

Management is normally reluctant to admit that their current system is inefficient and even if the top level management understands that it has a problem and makes a conscious decision to do something about it, juniors are often not prepared to divulge the necessary information to produce a solution. There can be good reasons for this such as the fear of being discovered as inefficient, or indeed the threat of possible loss of job once it has been proved that a machine can do the job more efficiently. There is also a reluctance to try anything new or to change procedures which are well understood by all the members of an organisation, however weak the organisation is known to be. The solution to this problem requires a very positive management approach. It must include, first of all, the education of all those involved to appreciate the capabilities and limitations of computers. Then, all members of the staff who will be involved must understand the development and procurement system for computers. Finally they must be enthused to play an active part.

Another problem has been that individuals in the computer industry have in many cases in the past held very narrow views. In the early days they were mainly engineers or salesmen, not trained to philosophise or think at a high enough level;

this, coupled with the fact that the demand for computers outpaced development and production, meant that good advice was not always readily available. There has also been a lack of adequate training in the analysis of problems and the design of solutions and a general lack of uniform standards of competence in the computer profession which is expanding so rapidly that it finds the recruitment of sufficient suitably trained staff impossible.

The introduction of ADP systems generally takes the form of a feasibility study followed by the design of a model which, when suitable, is implemented. Later the original design can be improved to incorporate new evolving requirements. Most users adopt this pattern and the use of consultants, 'system houses', and 'software houses' in all phases is becoming noticeably more common so satisfying the need to obtain professional, balanced views before embarking on a costly venture. The majority of costly failures have probably been due to the user not knowing what he really wanted coupled with the implementer's consequent inability to estimate accurately the size of the task in terms of cost and time before the implementation started; thus contracts were cast in terms that could not be too specific and all penalties caused by post contract date discussion or changes to the requirement had to be borne by the user.

Scientific Systems

Scientific systems are used mainly in research and development establishments, Universities, and scientific, engineering and the research and development departments of firms. There is the same tendency to batch processing of data as in the conventional role but with an increasing tendency to ON-LINE use and running of programs. On-line programming is a method of programming by which the programmer can write his programs seated at a terminal, such as a VDU, to input his orders, or STATEMENTS, directly to the computer.

The relevance of this capability is in the power it gives to the scientist or engineer to develop programs to the particular needs of his own research or development work. It is impossible in these systems to define in advance of installation exactly what functions will be performed and thus, as well as the on-line programming capability, large libraries of software and a large amount of redundancy or spare capacity are needed to give flexibility. The specification of these systems is normally very broadly based.

The use of a computer as a calculating machine is straightforward and it makes calculations possible in support of design or research which were not previously possible. There is an increasing use of computers in modelling and simulation of complex processes; this enables the scientist to answer the "What happens if... ?" type of question when applied to the operation of, say, a warehouse before it is built, or to the operation or loading of a complex radio network while the radio sets themselves are still in the design stage. Another rapidly expanding role for this type of system is that of computer aided design, where all known constraints in a particular field are stored in the computer together with all the experience of previous similar designs; a good example is the design of a motor car, where all this stored information can be used to provide a design environment as a starting point for the designer. Many man-hours of preliminary effort can be saved as a result.

There is a danger that too much attention will be paid to the results or information gained from this type of system. The problem of 'blind faith', or that the computer will be used 'because it is there', thus stifling original thought, is a real one but providing it is used as a tool this need not be the case.

It is impossible, as mentioned earlier, to design a single computer system for all scientific purposes due to the wide spectrum of its potential use. So a similar way of introducing it as used for the conventional systems is normal but, because the functions cannot be defined in such detail in advance, the feasibility study and model system design are often very much compressed.

There is a wide use of terminals such as VDUs, and graphics terminals are becoming very common as low cost graphic capabilities become available. There are also occasional needs for really powerful facilities which produce the requirement for the capability to off-load large jobs to larger machines such as those to be found in University networks. The wide spectrum of use also makes heavy demands on the control language, the sets of commands available to the user for the organisation and control of the programs and resources in the system, not to mention the operations staff themselves. It is often difficult to reconcile the requirements for batch work and a very fast response time for users of terminals and altogether scientific systems are far more complex than simple ADP systems. On the other hand, the users normally understand the capabilities of computers much better than the layman.

Communications Oriented Systems

It is very difficult to find a title to describe this group. They are all systems which depend upon an established communications systems to enable the transfer of data and information between various points. Speed of communications is often critical. Applications are in such fields as airline management and seat reservation, banking, management information, and command and control in military or police environments. They require large databases to which random access is possible; appropriate programs will draw data from the various data files and will update it. The data files may well be sited in stores at different locations and there may be a random occurrence of transactions: this is often referred to as a 'stochastic' occurrence. Inputs, too may originate at any one of the locations and it is clear that the communications networks form a vital part of each system.

Although these systems are normally based on well defined requirements and have specific functions they can be, by virtue of being large and widely distributed, complicated and very costly. They represent a large and important growth area, and in many cases are second generation systems for organisations which have gained experience of ADP using less complicated conventional systems as a first step. In many cases this has been caused by the evolution of new technology but there is a school of thought which considers that the adoption of an ADP system is best tackled in stages as the education of the user and the designer improves.

By distributing the input and output devices and some of the computing power around a network the system can be managed or controlled on the basis of timely, up-to-date rather than stale information. It is often said that "knowledge is power" but for this to be true the information on which the knowledge is based must be correct.

Any system which requires the processing of data which has been input to the system to produce a result virtually simultaneously is referred to as a REAL-TIME system. For example in an airline booking system, each booking must be processed by the system immediately the customer requests it so that the booking can be confirmed and an up-to-date view of the actual situation can be held in the computer at all times to be readily available for the next customer, possibly at another booking office.

The main problem, apart from cost, is that the design for such systems is extremely difficult while the penalties for making a mistake in the design are much greater than for a conventional system. The intractable design difficulty is to predict, accurately, the load on the system, even when a clear requirement has been stated. If the load is underestimated messages which include processing tasks will build up in queues much faster than they can be dealt with; then the system response time will probably increase to an unacceptable length and indeed the system may not be able to function at all. Thus there is a need for the designers to have a very high level of computer knowledge. It will involve an ability to analyse requirements accurately and an ability to use mathematical techniques such as simulation. An ability to translate the results into sensible design is also necessary and, finally, this ability must be tempered with a sound appreciation of cost effectiveness, for it is normally uneconomic to 'design for the peak'. It is not surprising that such people are difficult to find! The management of the development of these large, complex systems also presents many problems and this subject is dealt with in more detail in Chapters 3 and 4.

In the main, the input/output devices used are VDUs, teleprinters, and special-to purpose hand held or vehicle borne equipments, working through ranges of sophisticated communications processors which are controlled by intricate SYSTEMS software, the software designed to manage the operation of the total system. In contrast APPLICATIONS software, which can vary from very simple to very complicated, is the user's tool and can probably be loaded in, or unloaded from, the computer very easily. In a military context it might be possible to change a vehicle borne command and control system from one role to another simply by changing the applications programs. More important it should be possible to develop a single range of hardware and system software which can be tailored for a large number of users, thus enabling the advantages of commonality of equipment, in terms of lower cost and easier maintenance and training, to be realised. The training task must always be of prime consideration and it is in these time and situation critical systems that the interface between the user and the computer, known as the MAN/MACHINE INTERFACE (MMI), becomes tremendously important.

The method of introducing these systems is much the same as for conventional systems but, as they are often designed to individual taste and, in military systems in particular, to very specific standards, much time is spent in the evaluation of all possible development options.

Process Control Systems

The term process control systems is widely used in industry to describe systems which control plants such as refineries and steel mills, and includes modern microprocessor based systems, often part of some other larger system. The term is used here to include, in addition, military control systems.

An example is weapon control such as air defence radar and missiles, where a very rapid response to external stimuli is required. The computer system is very often in these cases only part of a larger system and inputs will come from other system components, often in an analogue form such as an electric current from a radar set or the radial position of a cog wheel in a machine. In weapon control, in particular, analogue computers have been used for many years and now the digital computer, often in the form of a microprocessor, is being introduced to give greater flexibility and precision. In many applications a computer may well be the only way of coping with a large amount of rapidly changing information which has to be processed. In air defence and air traffic control systems in particular, the ability to store and selectively display information is vital.

Once again, although it may be a relatively straightforward task to specify the function, the design is likely to be hampered by an inability to forecast a loading. In addition there may well be inputs and outputs which have to be dealt with at fixed times or the data may be lost or the elements being controlled will not function correctly. Unlike the communications oriented systems where it is costly, inefficient and therefore unusual to design for peak loads, it is essential to design process control systems for them.

It is in these systems that a range of almost philosophical questions is raised concerning the user's relationship with the computer: the main one is to what degree the computer can be allowed to make decisions. The answer needs a Solomon-like wisdom!

IMPLEMENTATION

In early systems implementation tended to follow a conventional pattern but experience has shown, especially in the military field, that the implementation of such systems has in many cases been costly, slow, and inefficient due mainly to the lack of a firm specification of requirements at an early stage, and a preoccupation with the costing and scaling of hardware before the full software implications of the total system function had been studied. This situation applies in all of the last three systems that have been considered where, after many years development, plenty of hardware is available but the software does not work correctly. Any large computer system involving a number of interrelated components must be considered from an overall system design point of view first and a software plan must be formulated. There will almost invariably be several hardware components which can meet the various hardware specifications identified by the software plan and it should therefore be possible to select those necessary to do the job. The problems of implementing the systems are dealt with in subsequent chapters.

SELF TEST QUESTIONS

QUESTION 1 What requirement did Babbage identify for a digital computer?

Answer ....

....

QUESTION 2 What is the main difference between analogue and digital computers in the way in which information is represented?

Answer ....

....

....

QUESTION 3 Define a 'system'.

Answer ....

....

QUESTION 4 Define a BIT.

Answer ....

....

QUESTION 5 Define a BYTE.

Answer ....

....

QUESTION 6 Define a WORD.

Answer ....

....

QUESTION 7 What are the three main components of a central processor?

Answer ....

....

QUESTION 8 What is the difference between 'direct' and 'serial' access store?

Answer ....

....

....

....

QUESTION 9 Define a PROGRAM.

Answer ....

....

QUESTION 10 What is the role of the control unit?

Answer ....

....

....

ANSWERS ON PAGE 197

2.
Hardware

INTRODUCTION

In Chapter 1 hardware was described as all the physical components of a computer system; that is all the electronic and magnetic devices. In this chapter these components are dealt with in more detail and their relationship with the operator or user is explored. As has been explained, digital computers have evolved from large mainframe static equipments to minicomputers providing portability for the first time and eventually microcomputers offering a tremendous explosion in possible applications.

CENTRAL PROCESSOR UNITS

Functions

In Chapter 1 it was explained that a CPU, consisting of a control unit, ALU and some associated store, called main store, carries out the basic function of a computer, which is the processing of data. The control unit controls the logical operations of the machine and causes them to be carried out in the correct sequence while the arithmetic unit carries out arithmetic operations on the operands. The physical size of CPUs has reduced dramatically in the progression from mainframes, to minicomputers, to microcomputers.

Speed

There is a constant requirement to increase processing speeds and one relatively simple technique is overlapped instruction execution, or PIPELINING, where a processor while executing one instruction, will call forward the next instruction at the same time in order not to have a gap between instructions.

Another method of speeding up processing is to employ more than one processor. By the use of several special purpose processors it is possible to allocate concurrent tasks to different parts of the system. One example is the use of a special input/output processor to handle peripherals; another is a special processor to carry out floating point arithmetic: such a unit can be incorporated on a single chip. The use of more than one processor also allows concurrent instruction execution or PARALLEL PROCESSING. A chart showing a comparison of size with processing power is at Fig. 2.1.

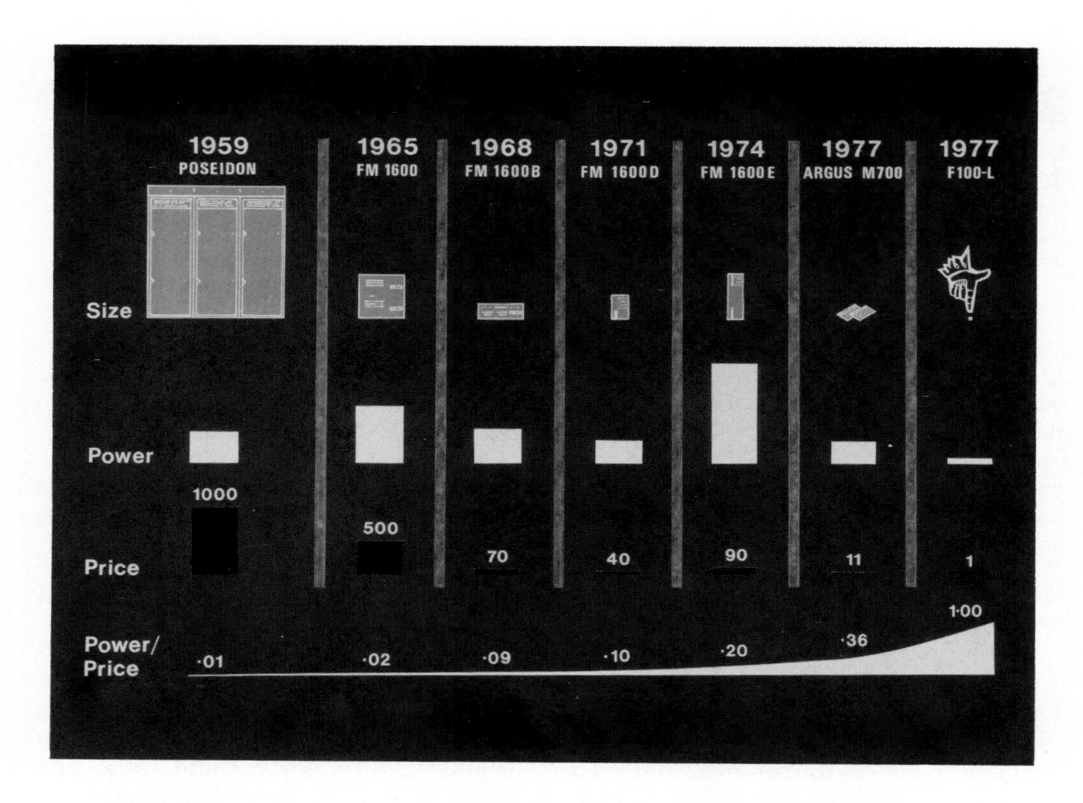

Fig. 2.1 Comparison of size and processing power

Reliability

There are several techniques evolving in order to provide better reliability. For instance, redundant self-checking auxiliary processor units to check results at any level, inter-processor CHECK POINTS to provide snapshots of a situation for recovery purposes in the event of failure, and KEEP-ALIVE and automatic restart facilities for use in the event of power supply failure, are quite common.

Effectiveness

Exception-handling measures can provide for the occasions when a processor will perform to its specification but results are not acceptable to the user, possibly due to software errors. For example a processor may be tasked to multiply two alphabetic characters; it would carry out the process but would indicate to the user that there had been a problem.

Techniques to provide the overlapping of instruction execution and input/output, allow input and output devices to access stores while the CPU is carrying out other instructions; a common term used here is DIRECT MEMORY ACCESS (DMA) which is discussed further in Chapter 3.

One way of designing suitable permanent INSTRUCTION SETS, the interfaces between the machine and machine code, is called HARD MICRO-PROGRAMMING. SOFT MICRO-PROGRAMMING on the other hand, provides for continuous manipulation of the instruction sets; in other words it enables the MICROCODE, the internal software which determines and defines the user machine code instruction set, to be changed to match changes in high level languages. These facilities are becoming common in new systems. Figure 2.2 shows a complete packaged microprocessor.

Fig. 2.2 MC 1800 microprocessor

STORAGE DEVICES

The digital computer is based on the stored program principle and thus depends upon the availability of efficient storage devices to store both the programs and the data. Whilst both input and output devices and CPUs have undergone rapid and exciting development, storage systems have not changed in concept and remain a constraint on design and structure. A storage system may need to have extremely efficient fast mechanisms for storing data to provide a fast reaction time for the user but at the same time it must make efficient use of its capacity, for store is expensive because of the manufacturing processes involved. Storage devices are employed in two ways; either as main store or backing store. (The latter is sometimes referred to as BULK or MASS memory).

MAIN STORES

Functions

The main store is direct access memory and has two functions. The first is to provide very fast devices with active elements, such as integrated circuits, for use as local stores or stores to hold temporary results (known as SCRATCH PAD memory) and for control and arithmetic operations. They are associated, for instance, with registers to store intermediate calculations and work closely with the CPU. They could equally be ROM employed to control and execute repetitive instructions or hold sets of tables.

The second function is to hold programs being executed, and all the current data. For this, RAM of large capacity will be necessary.

Ferrite Cores

To date mainly ferrite cores, thin magnetic film, or plated wire have been used to construct RAM storage devices; ferrite cores were popular because of their comparative low cost, but are bulky and are thus less portable. The low cost and high packing density of semiconductor RAM means that it is now the preferred method in most new designs.

The principle of ferrite core depends on a small ring of magnetic material (ferrite is a magnetic oxide of iron) which can be magnetised in either of two directions. A clockwise polarity can be considered as a 1 and anti-clockwise as a 0 (see Fig. 2.3). Once current has been passed down wires and the core is magnetised, the current can be removed and the magnetism and polarity will remain. This is the way in which digits are read in.

To read data out, a current of known direction is fed through the wires; if the core was previously magnetised by a current flowing in the same direction then no flux change will occur. However if the known current is different to that which had been applied a reversal of the magnetic flux of the core will occur and a pulse of current will be fed down a SENSE wire which is also threaded through the core. In order to read the data held by the core the bit was driven-out and this process is known as DESTRUCTIVE READ-OUT; in order not to lose that data it has to be rewritten into the core, necessitating the use of yet another wire.

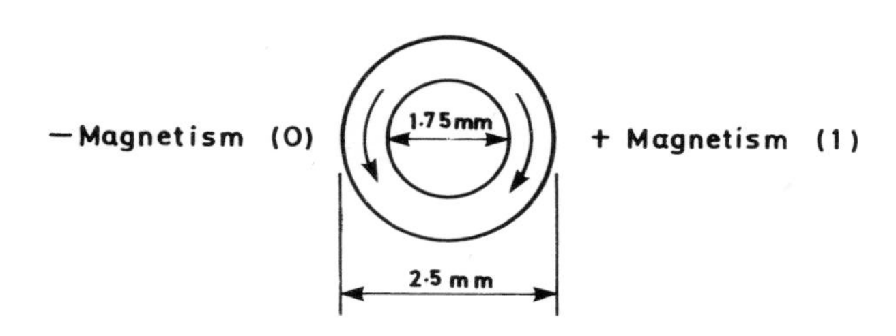

Fig. 2.3 Plan view of a core

A typical arrangement of cores is shown in Fig. 2.4. The cores are threaded onto wire grids at the points where the X and Y axis wires cross. The polarity change for each core is effected by passing a current along the two wires which cross at that core. The sense wires for read-out are interlaced diagonally.

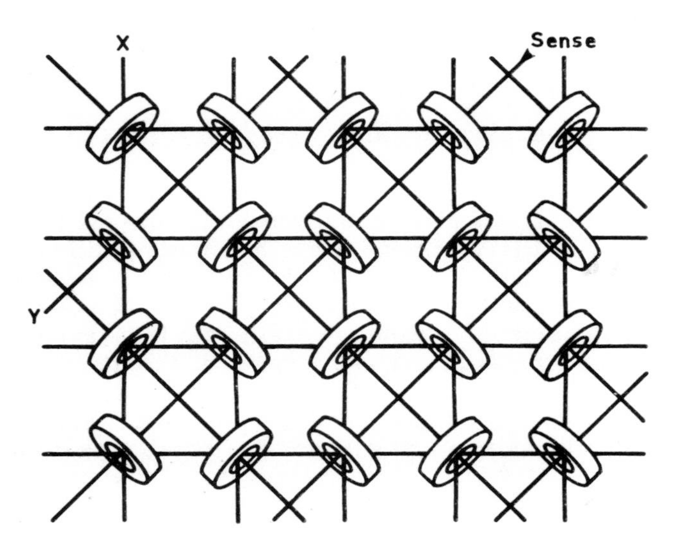

Fig. 2.4 Part of one core store plane

In practice each core has to be threaded by hand and is costly to assemble. Each core must, of course, be large enough to permit threading and this, too, adds to the limitations on the use of core store.

Magnetic Film

Data can also be stored in magnetic form on a thin magnetic film deposited on a plane surface, a bar or a wire. The direction of magnetisation of a small area of the film determines whether a 0 or a 1 is stored. This data is changed or read by passing current along appropriate wires which address each area. It was originally forecast that very large capacity magnetic film would become widely used to replace many of the backing storage devices in current use but this potential has not yet been realised and it is doubtful whether it ever will.

Semiconductor Store

It is now possible to manufacture simple electronic circuits which can exist in one of two stable states (bistables). By using one state to represent a 1, and the other a 0, data can be stored in them. The development of integrated circuit techniques allows tens of thousands of such circuits to be implemented on a single silicon chip. It is therefore clearly possible to produce a main memory of considerable capacity within a small physical space. A distinct disadvantage, however, is that semiconductor store is VOLATILE in that, if the power supply is removed, the data is lost.

BACKING STORES

Functions and Types

Backing stores are used to store additional programs that are not being run currently, and data that must be reasonably quickly available to go to the main memory. There is a need, therefore, for high capacity, medium speed but low cost devices. Magnetic drums and discs are used traditionally in this role but other devices such as BUBBLE MEMORY are becoming available. All these are explained later. When speed of access is not important it is possible to use magnetic tape to store large quantities of data very cheaply; this is particularly useful when, for instance, there is a vast amount of data to be archived. The backing store is organised into independent modules in order that files such as discs or tapes full of data can be removed from the system for storage or for use in other computer systems. In drums, discs, and tapes the data is stored on the magnetic surface with the direction of magnetism in a small area determining whether a 1 or 0 is stored.

Magnetic Drum

On a magnetic drum, which is now becoming a rather outdated device, information is recorded in tracks round the surface with a read/write head for each track. In some drums the read/write heads float on a cushion of air to keep them out of contact with the drum surface. The drum is also very slightly tapered and moves down, when not being used, to avoid the read/write heads touching the surface of the drum as shown in Fig. 2.5.

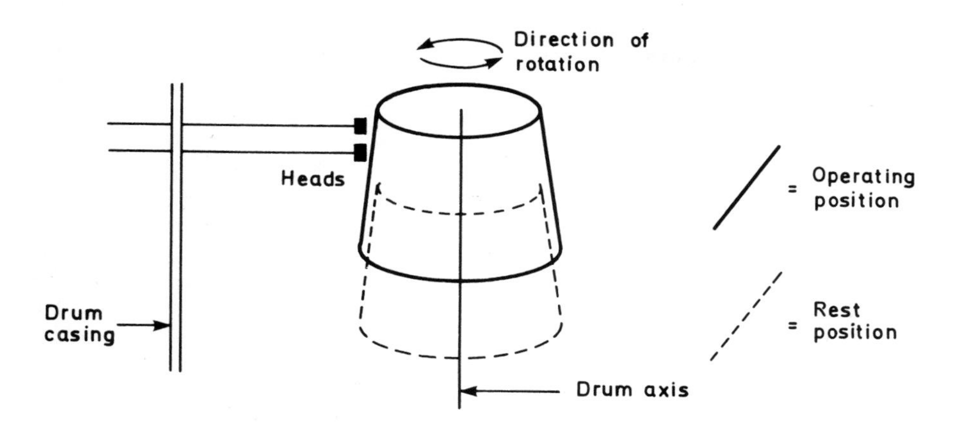

Fig. 2.5 Representation of a magnetic drum

The plan of the storage of data on a drum is shown in Fig. 2.6. A typical drum can store 4000 characters on each track, has a total capacity of 5,000,000 characters and can transfer data at the rate of 1,300,000 characters per second.

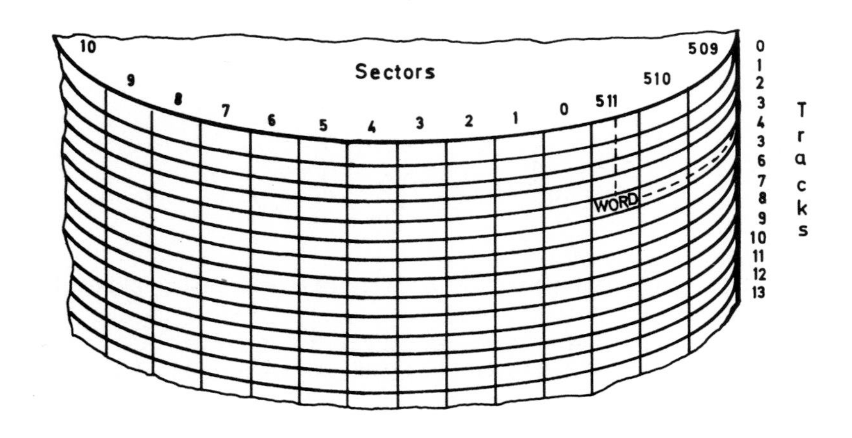

Fig. 2.6 Addressing on a magnetic drum

Exchangeable Disc Store (EDS)

The main functional componęnts of an exchangeable disc store (EDS) are the CARTRIDGE and the TRANSPORT, which contains it. A typical cartridge consists of six recording discs permanently mounted on a vertical rotating shaft. Each disc is made of metal, coated with a thin layer of iron oxide, typically 1.27 mm thick by 356 mm diameter. One recording surface consists of some 200 concentric tracks along which data is recorded by read/write heads. There are 10 recording surfaces, the upper side of the bottom disc, the lower side of the top disc and both sides of the remaining four discs.

The main operational features of the transport are the read/write heads, which are mounted on retractable arms. When the arms are retracted, the cartridge may be inserted or removed from the transport. When the transport is started, the cartridge is accelerated to 2400 rpm after which the arms advance and the heads go to the FLYING POSITION floating on a thin film of air about .025 mm from the disc surface. The head assembly is shown in Fig. 2.7. One EDS Control Unit can easily control up to 8 transports but simultaneous access to two or more transports may not be possible.

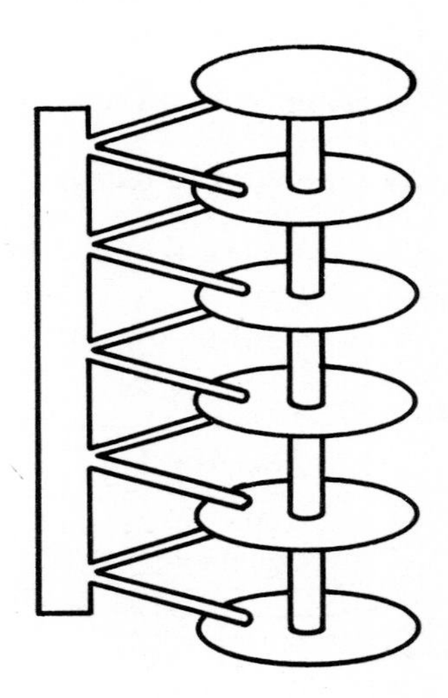

Fig. 2.7 Position of magnetic disc heads in operation

Data is arranged in blocks of words and there are several blocks to each track. Blocks are grouped into SEEK AREAS which can be accessed with the heads remaining in the same position. Addressing, as in the drum, is by hardware: addresses can easily be designated according to block number and position of the heads. A typical EDS can hold over 67,000,000 characters and can transfer data at over 1,000,000 characters per second.

Magnetic Tape

Magnetic tape consists of a flexible plastic base coated with a binding medium containing microscopic particules of iron oxide. It is in this ferromagnetic coating that recording occurs. The width of tape is approx 12.7 mm and it is wound on reels which are available in a number of sizes. The largest reel is 266.7 mm in diameter and holds about 700 m of tape.

The main components are tape spools, vacuum chambers containing a reservoir of tape which enables it to be accelerated or decelerated rapidly, capstans, pinch rollers, and finally read, write, and erase heads as shown in Fig. 2.8.

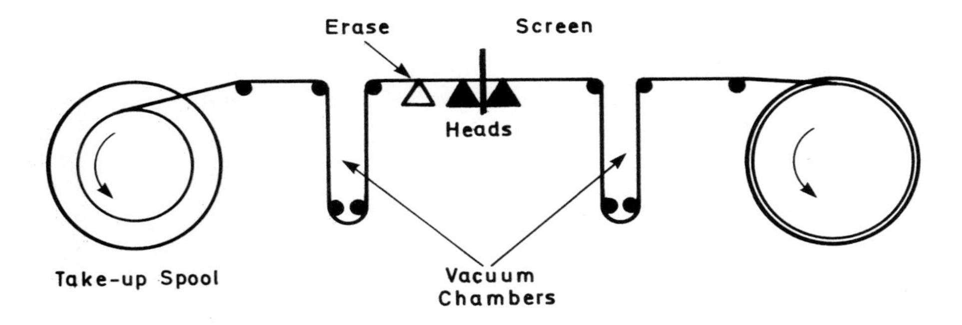

Fig. 2.8 Layout of tape deck components

The read and write heads each consist of a ferromagnetic core around which are wound coils of wire; the two heads are separated by a centre screen. Current pulses in the appropriate direction in the write coils induce either a 1 or a 0 in the ferromagnetic surface of the tapes and the reverse process occurs during reading when a small but detectable voltage is induced in the read wires as a magnetised area of tape passes across the read head. The tape must of course be moving in the correct direction at the correct speed before reading or writing can take place. The erase head is located in front of the read and write heads and erases the obsolete information on the tape before new information is written.

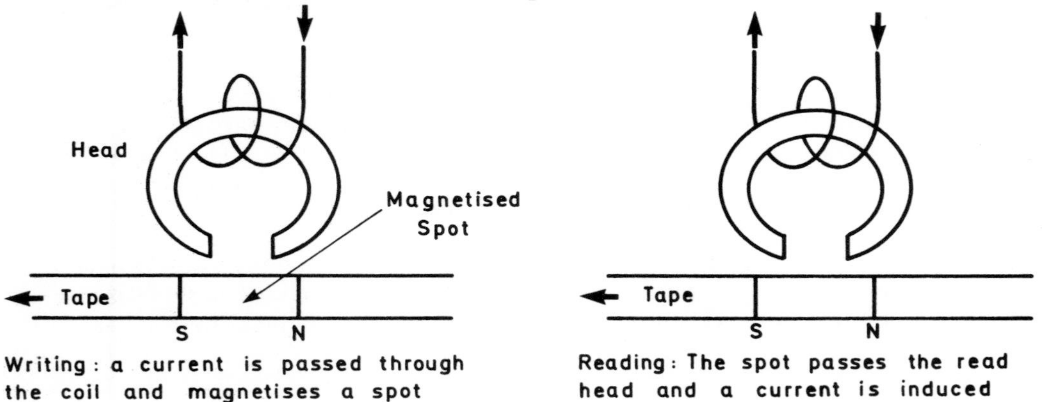

Fig. 2.9 Writing and reading on magnetic tape

Data is stored in blocks, with gaps between blocks where the tape can be accelerated, and is transferred to and from the normal backing stores block-by-block. The complete tape is read to enable addressing to be carried out and each file name is checked as it passes under the read head until the correct file name is found. Tapes hold between 200,000 and 40,000,000 characters which can be transferred at between 25,000 and 40,000 characters per second.

Floppy Disc

An interesting development, from a military point of view, is that of the FLOPPY DISC where recordings are made on a single oxide coated plastic disc, similar to a 45 rpm record. The recording head is placed in contact with the surface when a read or write operation takes place and the wear this causes means that it is not good for long term use. Its applications are in small, compact, environments which do not demand high capacity or high transfer rates. The disc rotates at 360 rpm; it is cheap, easily handled, and normally holds 500,000 characters which can be used at a rate of 30,000 characters per second (see Fig. 2.15).

Bubble Memory

A recent interesting development, because it is small and has no moving parts, is BUBBLE MEMORY. It is manufactured by depositing a thin garnet film containing iron onto a nonmagnetic garnet substrate. The iron atoms combine with atoms of other elements in the film and behave like small magnets. Bonding forces in the garnet crystals cause these combined atoms to align themselves in parallel with their closest neighbouring atoms; in this way magnetic regions or DOMAINS form in the film whenever groups of atoms align themselves in the same direction.

The application of an external magnetic field expands all domains with the same polarity as the field and causes all others to shrink. The presence of a bubble represents a 1 and the absence a 0. Once bubbles have been formed, the application of a constant magnetic field will maintain them so a simple magnet placed next to a bubble memory chip will preserve data without any further power being applied. Once a bubble is created it is shuttled around on the chip, to a specific area for storage, along paths of soft nickel-iron material. Since bubble movement involves only the magnetic field domains moving through the crystal and does not involve solid matter, the moves can be effected very rapidly. The bubbles are read by passing them in front of a detector which senses them and converts them to data, ie 1s and 0s, but the data will, of course, be read out serially.

Modern chips can hold up to 128,000 characters with a transfer rate of 8,000 characters per second. Figure 2.10 shows a card of 16 bubble memories on their chips produced for the UK command and control system WAVELL.

Fig. 2.10 Bubble memory card

Cassette Tapes

Open reel tapes are often used for large-scale data processing on a mainframe computer, whereas cassette tapes find applications in smaller systems which do not need such high transfer rates or capacity. They are cheap devices and need less skilled operators; applications include holding libraries of programs and databases with access times of several seconds or even minutes. There is a wide range of such devices available and their characteristics are 10, 000 to 160, 000 characters per tape with transfer rates between 60 and 1, 200 characters per second.

Balancing Store Requirements

The main qualities required of any storage technique are rapid access to data, cheapness and robustness. These qualities cannot be provided in parallel for devices providing rapid access to data, and the most robust, are the most expensive. Thus the hierarchy of storage devices described in this chapter has evolved; fast and expensive techniques are employed in the hard working register stores holding small quantities of data for the CPU, and cheaper and less readily accessible stores are used to accommodate data to which less frequent access is required.

INPUT AND OUTPUT DEVICES

Functions

In broad terms, as we saw in the first chapter, input and output devices serve as the communications link between the user and the computer. They are controlled by some form of input/output unit which, in its turn, serves as the communication between the CPU and all its peripherals. The main task of the unit is to effect transfers of data between a high speed CPU and the much slower running external peripherals; to do this there is a good deal of data preparation to be carried out. Many types of input and output device can be used in conjunction with a CPU and several of them with possible military application are described below. One special device that must be mentioned first is the analogue to digital converter. It is necessary to convert analogue data from transducers and measuring devices into digital signals in such applications as real-time control systems; the reverse facility must also be included.

Punched Tape Input

Punched tape is the cheapest and most commonly used form of input. Tapes are normally up to some 30m long and the characters are read in at speeds varying between 10 - 1000 groups per second. Programs or data are put on a paper mylar tape by punching holes in groups of 5 or 8 in a prearranged code. (See Fig. 2.11).

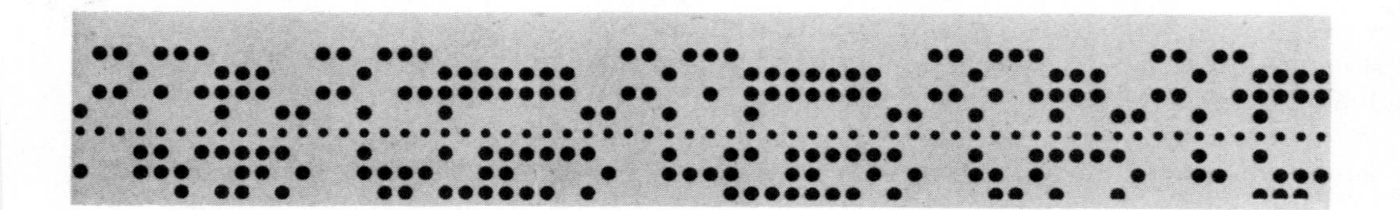

Fig. 2.11 Punched paper tape

The holes are sensed by the reader unit mechanically, photoelectrically or dielectrically and the tape is fed through the reader electromechanically.

Punched Card Input

Punched card is widely used in large data processing installations where there is time for off-line preparation. Holes are punched in cards in 80 columns in 12 rows and are characteristically round or rectangular. The holes are sensed photoelectrically or electromechanically, serially one column or row at a time. A magazine holds between 250 - 3500 cards and a card reader can read between 250 - 800 cards per minute. An example is at Fig. 2.12.

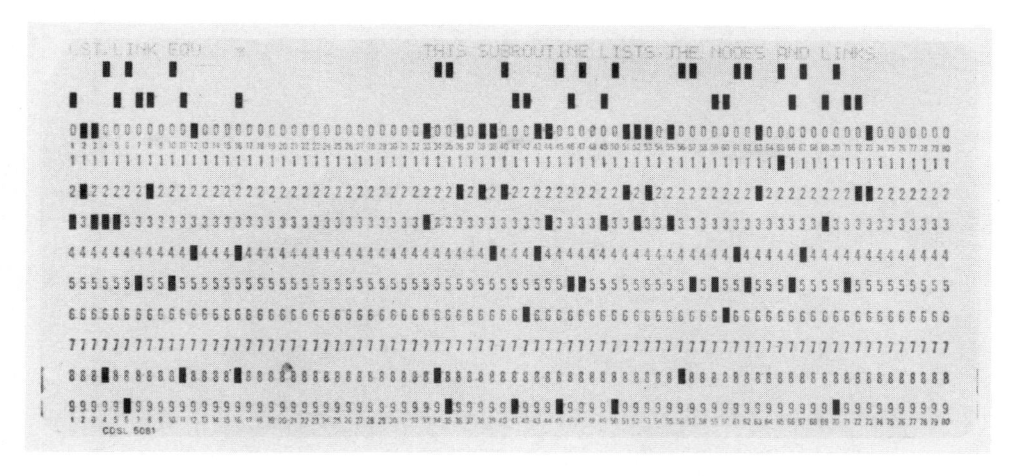

Fig. 2.12 Part of a punched card

Graphic Inputs

A cathode ray tube (CRT) can be used to provide a graphic display and a 'light pen' can be used to identify points, to draw extra lines, or to amend lines already on the display. Photoelectric scanning is used to provide these interactive capabilities so that the operator is in direct communication with the computer and is able to obtain immediate response to his input messages. A common application is that of COMPUTER AIDED DESIGN (CAD). The capability may also be appropriate for use in interactive map displays of the future. A powerful new tool is the GRAPHIC TABLET, which allows an operator to draw on a sheet of paper placed on a special tablet using a special pen, and the digitised position of the lines drawn are then fed into the computer automatically.

Teletypes

Teletypes incorporating a punched paper tape reader, printer, and with a standard alphanumeric keyboard or with special to purpose keyboards are fairly common, especially in real-time systems. Electromechanical sensing is normally used but photoelectric means are becoming popular. The keyboard generates bit codes but can only produce 10 - 15 characters per second at maximum.

Digital Message Terminals

DIGITAL MESSAGE TERMINALS are normally microprocessor based and designed for use with tactical radio sets. They provide much more efficient use of the available radio frequencies by enabling messages to be transmitted digitally in a fraction of the time taken to send a similar verbal message and thus maintain accurate communications in hostile environments. A message is prepared on a keyboard, possibly displayed to the operator at the same time, and then transmitted in a burst via radio to a receiving station. A typical application is for the transmission of complex artillery fire orders which can be achieved more rapidly and accurately, and such an example is shown in Fig. 2.13.

Fig. 2.13 The US Army Digital Message Device (DMD) in service as part of the TACFIRE system. (Other, general purpose, devices are illustrated on the cover of this book and in Chapter 7)

The addition of an encryption device would mean that, as well as dramatically reducing overall transmission time and improving accuracy, security against information being compromised is provided.

General Output Device

Paper tape and punched cards are also used commonly as output devices from large, static computers. Teleprinters and teletypes are widely used in a variety of applications; again they are slow but can be extremely useful as system monitoring or control devices (see Fig. 2.14). They produce printed copy, and some equipment is capable of extremely high quality printing which may be used directly for producing documents. High speed, high volume line printers are very fast but very expensive. A useful additional facility can be the employment of a GRAPH PLOTTER, which accepts electrical output signals from the computer and prints these in a pictorial or graphical form on paper or transparent material.

Fig. 2.14 A militarised, teleprinter style, hard copy terminal

Visual Display Outputs

The problems of keeping the volume of information available to military users within manageable proportions have focused attention on the manner in which the information is presented and a common feature of evolving systems is the use of visual display sub systems. These could consist of a display processor which controls display drives for cathode ray tube displays, large screen displays, and gas discharge PLASMA PANELS. Modern developments in these areas rely particularly heavily on the use of microprocessors which allow a standard set of devices to be reconfigured in different ways and for new features to be added at short notice, because the display device requirement will vary from user to user.

Using a microprocessor it is possible to reconfigure the display device control by changing a plug-in memory chip.

VDU tabular displays are commonly used in systems where information needs to be presented quickly to an operator. Alphanumeric characters are commonly displayed at speeds of 1,000 characters per second and common displays will hold 20 lines at 80, or sometimes 132, characters per line. An example of a VDU is shown in Fig. 2.15 together with a keyboard and floppy disc drive.

Fig. 2.15 A typical VDU, keyboard and floppy disc drive

In military environments conventional CRT VDUs can be too bulky, unreliable, and have too short a life. Plasma terminals are now available, using a 'glass' which is in fact neon gas sandwiched between two plates of glass each with a grid of electrodes attached to the inner surface. A 'driver and sustainer' generates the voltages to be applied and data is passed along the display axes to illuminate dots where applicable; a common display has 512 dots per line. Plasma displays offer the advantage of being flicker-free and have a longer life expectancy than the traditional CRT, but they are currently more expensive. It has been suggested, however, that if the initial purchase and in-service costs of the two devices are compared there is little to choose between them.

Many users can identify the need for graphical displays on VDUs, whether CRT or plasma panel. The applications range from displays of bar charts for management purposes to nuclear fallout prediction diagrams and weapon coverage traces. Providing the hardware has been designed to have sufficient INTELLIGENCE, which means processing capability, software to provide graphic presentation of information is not difficult to produce.

ERROR DETECTION AND CORRECTION (EDC)

The accuracy of digital transmission depends upon the use of error detection and correction techniques; this is most important in battlefield HF communications in which fading can cause problems. Error detection codes check characters against specific rules and reject mis-matches. Due to the low power consumption of current integrated circuit technology it is possible to build an error correcting coder and decoder device on a printed circuit board. Efficient examples, that is to say devices which keep the transmission rate as fast as possible at the same time as measuring and correcting errors, are now available.

MAINTENANCE

Built-in Test Equipment (BITE)

BUILT-IN TEST EQUIPMENT (BITE) can enable many faults to be detected and diagnosed under field conditions, whereupon a unit may be exchanged or a plug-in printed circuit board replaced. More detailed testing can then be carried out at base workshops, often using AUTOMATIC TEST EQUIPMENT (ATE). The low cost of the hardware may often make it more economical to throw away a circuit board rather than try to trace the fault to a particular component and then repair the board.

Automatic Test Equipment (ATE)

The concept of ATE evolved some years ago when the complexity of military electronic systems became too much of a burden for conventional maintenance methods. Until that time most military projects involved single suppliers who built complex ATE for their own products. Today the size of systems means that more than one supplier provides component parts and thus the trend is towards general purpose ATE for military use. Even general purpose ATE must allow either diagnostic testing and fault finding down to component level, or functional testing to module level.

Before ATE was commonplace, the normal repair routine was to replace modules; each module would consist of several sub-units or boards which might have been operational apart from one fault. Obviously it was costly to keep large numbers of complete spare modules in store, and, as turn around times for equipments undergoing repair is critical, ATE evolved as a solution; the choice between functional module testing or dynamic board testing being determined by the type of equipment under test or the condition of test.

Radio, radar, and large ADP system modules make functional testing important, because accurate testing can only be carried out in operational scenarios. This will necessitate complex test equipment that must also cope with simple board testing which, although probably involving only 'go/no-go' routines will be vital in forward repair units where detailed repair is impossible.

MAN/MACHINE INTERFACE (MMI)

The Need

The digital computer was first introduced into scientific and then into commercial and administrative work. It was appreciated very early on that one of the main difficulties in using computers was communicating with them. Often the only language available was the computer's own language and this meant that the user had to work through one of the programmers capable of talking to the computer in this language. Later high level languages became available and more sophisticated computers and operating systems reduced the turn-round time but the situation was still basically the same: the user was working at some distance, in space and time, from the means whereby his problem was solved.

The disparity in speed between man and computer also meant that it was cost effective to have several users working simultaneously. In certain fields, notably scientific and technical work, demands arose for more direct access to the computer and ON-LINE working evolved. This is explained in more detail in Chapter 3.

The Machine

The history of the relationship between man and machines has seen the machines steadily rising in status relative to man. At first machines were very simple and for the most part had to work under man's direct control (eg the cart and the winch). Fairly early on machines were developed which could work independently of man, the windmill and waterwheel for example, but only on very simple tasks. The first machine to achieve a degree of independence on a relatively sophisticated task was probably the programmable silk loom invented by Jacquard as long ago as the 18th century.

Digital Computers

It is sometimes said that the digital computer is the first extension of man's brain as opposed to his limbs. It is the digital computer's ability to extend towards higher level capabilities such as aiding decision making, complex calculation and storing data that differentiates it from other machines. The digital computer has grown steadily in intellectual and executive status since its introduction, until now it is working almost as an equal partner to man in various fields. In the early 1960s it was thought that this process would continue at an increasing rate and remarks such as "Man must always think he is in control" were made: at that time it was envisaged that effective control would pass to computers. A somewhat more prosaic view prevails now, but the development and the consolidation of the computer's increasing status is likely to continue.

Man Versus Computer

While generalisation is often inappropriate, it is useful to compare some of the characteristics of man and the computer in order to determine the optimum mix of tasks allocated to each of them. Man tends to be good at pattern recognition and has a vast associative memory which may not, however, be very accurate for details. He has a very flexible input mechanism and he is capable of making decisions based on experience, judgement and intuition even using imperfect information.

The computer, in contrast, has a very reliable and detailed memory but requires exact or nearly exact keys for pattern recognition. It is capable of rapid calculation, analysis and fast retrieval with a very fast and flexible output. It can also make excellent decisions providing it is fed with perfect data and a fool-proof decision making program.

Relationships

The relationship between man and the computer has many levels and some examples, in ascending order of the computer's relative status, of what the computer can do to assist man are:-

- Output the basic data that it has stored
- Output information after it has processed the data
- Co-operate in problem solving
- Provide guidance and suggestions
- Share control
- Take full control and report back progress.

The majority of computers are employed in the first two categories, and thereafter some joint activity is necessary. Shared control occurs when the computer is allowed to free-run for much of the time only informing the man when assistance is required. When the computer assumes full control, as in some guided weapon systems, man only requires that the computer informs him of the progress made.

Implementation of MMI

MMI involves software, hardware, and environment. The military environment is severe. Such factors as climate, protection, time and mobility manifest themselves in many constraints such as protective clothing, restricted space, and tiredness. All military weapons or instruments must take account of this in their design, because unless a soldier can use his equipment efficiently he will not be effective. Man's relationship with the computer can be described by considering methods of communication, levels of user, and the direction of the communication.

Consider man communicating with the computer. The two main functions in this case are to give commands and to request information. There are three levels of user; the programmer, the skilled operator, and the lay operator. The programmer has several levels of computer language available to him, each with its own benefits and drawbacks, from which he can select the most suitable. The skilled, or expert, operator is trained for a special function and his communication with the computer may be by means of sophisticated retrieval languages, special purpose shorthand or light pens, touch wires, joysticks and other manual devices. Lay operators, who might be expected to know roughly how a terminal works, but little more, must have specific guidance via mechanisms by which they can be led, step by step, through a conversation with the computer.

There are also many ways in which the computer can communicate with man, as demonstrated by the large variety of hardware output devices available, and these, of course, match man's exceptionally flexible input characteristics. There are, however, two distinct functions in computer-to-man communication. The first is the message function in which messages or reports are generated, normally in reply to input messages, and the second is the commentary function which is normally only achieved via output devices, not involving paper, which are very fast: a VDU is a good example. This latter function introduces a number of questions such as the selection of items to be displayed without saturating the man with information. Very careful analysis of the man's function and the level of man/computer relationship is necessary. If the computer is being delegated a good degree of control, more information may be output than when the man is required to make more decisions; in such a case he requires only to have information that is strictly relevant to those decisions.

Response Times

Response times are closely associated with data structures, quantities of data held, levels of relationship and the pattern of system loading which, of course, affects message arrival rates. In general terms on-line peripherals, without their own associated store, cannot wait. On the other hand, man and other devices with a store can wait providing the system is suitably designed to allow queueing and storage of messages and data.

The optimum response time is approximately one to three seconds for on-line conversations. Any lengthening of this tends to lead to impatience and a reduction in fluency; but one simple aid to ease such a problem is the generation of a message to say that the particular processing called for is in fact in hand. This at least gives the man confidence and assuages his impatience. It is essential of course that military systems are designed for operation in peak load conditions.

THE WAY FORWARD

Hardware technology has developed at a tremendous rate over the past two decades and it will probably continue to do so. However it has been limited by the software which has lagged behind in development; the situation is now improving as we shall see in the next chapter.

SELF TEST QUESTIONS

QUESTION 1 What does 'volatile' mean when applied to memory?

Answer ....

....

QUESTION 2 What is the major difference between ROM and RAM?

Answer ....

....

QUESTION 3 What factors have to be taken into account when considering the options available in the selection of a storage device?

Answer ....

....

QUESTION 4 If a CPU is termed "8-bit", what does this mean?

Answer ....

....

....

....

QUESTION 5 Place the following types of store in ascending order of access time: a. semiconductor. b. exchangeable discs. c. floppy discs. d. bubble memory. e. magnetic tape.

Answer ....

....

QUESTION 6 What are advantages of bubble memory when used in military systems?

Answer ....

....

QUESTION 7 Why might core store sometimes be preferable to semiconductor RAM?

Answer ....

ANSWERS ON PAGE 197

3.
Software

INTRODUCTION

Definition

In a general sense, software is the term used to describe all the programs associated with a particular computer and all the documentation associated with the development and design of those programs. In certain instances the term is used to describe only programs written for specific user applications. It is obviously necessary to confirm the local meaning of the term before using it. A large amount of general purpose software is normally provided by computer manufacturers and the developers of special-to-purpose systems can very often use sections of it to reduce considerably the effort required to design their particular system and the amount of new program writing required to implement the design. The size and complexity of software packages varies considerably: the basic unit is the SUB-ROUTINE.

Sub-Routines

Sub-routines are self contained parts of a program which can be incorporated into complete programs as required. They are useful because certain routines are frequently common to many large programs and very often a single program may wish to repeat a certain routine many times, particularly in mathematical functions; thus it is cost effective to write the routine once only. Another advantage is that once the particular piece of program in the sub-routine has been written and tested it becomes a standard and will thus make further programming quicker and easier.

COMPUTER LANGUAGES

The Need

In Chapter 2 we saw that the computer is a tool, an information system, an aid to decision making and can perform many other roles, whether employed as a single unit or as part of a system. It is in fact an extremely versatile tool which can deal with a vast array of problems and situations. We also saw that man needs to communicate with the computer in much the same way as he talks to another man. Since the computer was invented, designers have expended a great deal of effort in attempting to arrange such an interface. It is not an easy task for, as was mentioned in Chapter 1, the digital computer operates on binary code.

Low Level Languages

In early computers the programming had to be carried out in machine code but it is difficult for the average human to think in binary terms, and very inefficient. Consequently when it became apparent that separate computer programs could be used to do much of the detailed translation work, ASSEMBLY LANGUAGES using symbols were developed to replace numeric machine codes: these use mnemonic codes such as ADD for 'add' and SUB for 'subtract' functions. The programs developed to convert the assembly language source program written by the programmer into the machine code equivalent, or OBJECT program, are termed ASSEMBLERS. Assembly language is not, however, very far removed from machine code, and such languages are termed 'low level'. There is widespread use of them, but they remain closely allied to machine code. They can normally be used only on a particular machine, and the purpose of the program is far from evident to anyone other than the original programmer. Thus a means of translating from normal English, which is a set of phrases and their interpretations, to instructions which the computer could understand was still sought. This led to the development of higher level languages.

Autocode

The success of assembly languages encouraged the development of even higher level programming languages which, by design, achieve larger amounts of actual processing for each instruction written by the programmer. This obviously reduces the number of mistakes made by the programmer and makes it easier for others to read, understand and check the programs. Each high level instruction can equate to, say, five or more instructions in machine code.

The earliest of the higher level languages were called AUTOMATIC PROGRAMMING CODES or AUTOCODE and were developed almost exclusively for mathematical work. A good example is the simple Elliot Automation autocode produced for use with Elliot's range of scientific computers. Using Elliot autocode it is possible to add two quantities (say A and B) together to make a sum (C) by writing to the computer in straightforward terms: C = A + B

Similarly for other functions, C = A - B, C = A / B

These statements are of course mnemonic, in that they use well known algebraic notations, and also absolve the programmer from specifying in which store A, B and C are to be held the program will do this automatically. There are many limitations in the use of autocode. For instance, only capital letters may be used and there are many local rules, but, on the other hand, it is possible to learn autocode in a few hours sufficiently well to write simple programs. The greatest problem is that machine codes, assembly languages and autocodes are in the main peculiar to a particular computer or to a few models in one manufacturer's range; so it is necessary for a programmer working in languages at this level to learn a number of languages and it has been difficult for programmers working on different machines to converse sensibly and exchange ideas. The next logical step, therefore, was to attempt to standardise languages at the same time as improving them by making them even more flexible and higher level.

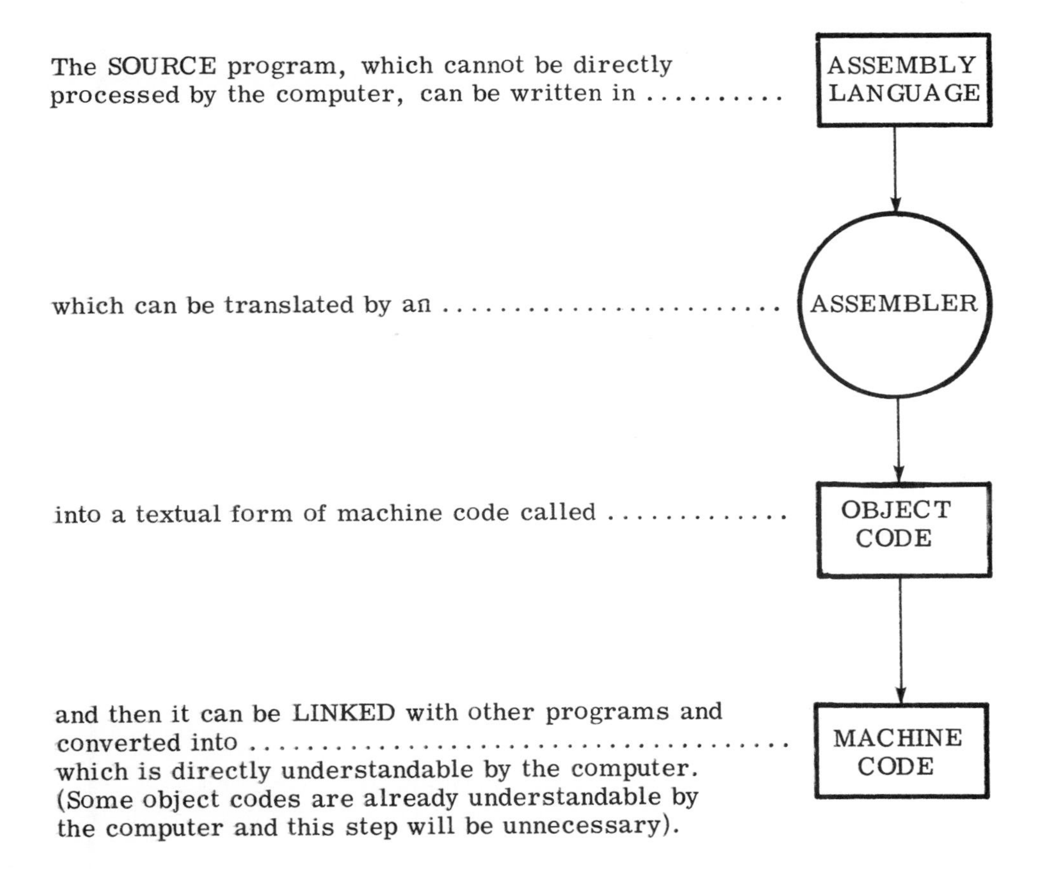

Fig. 3.1 Assembly language

The Development of High Level Languages

The aims of developing a high level language are to produce a language in which each instruction corresponds to several machine code instructions, and where the programmer can write using notations with which he is familiar. The translation is usually carried out by a COMPILER which will translate statements into many machine instructions.

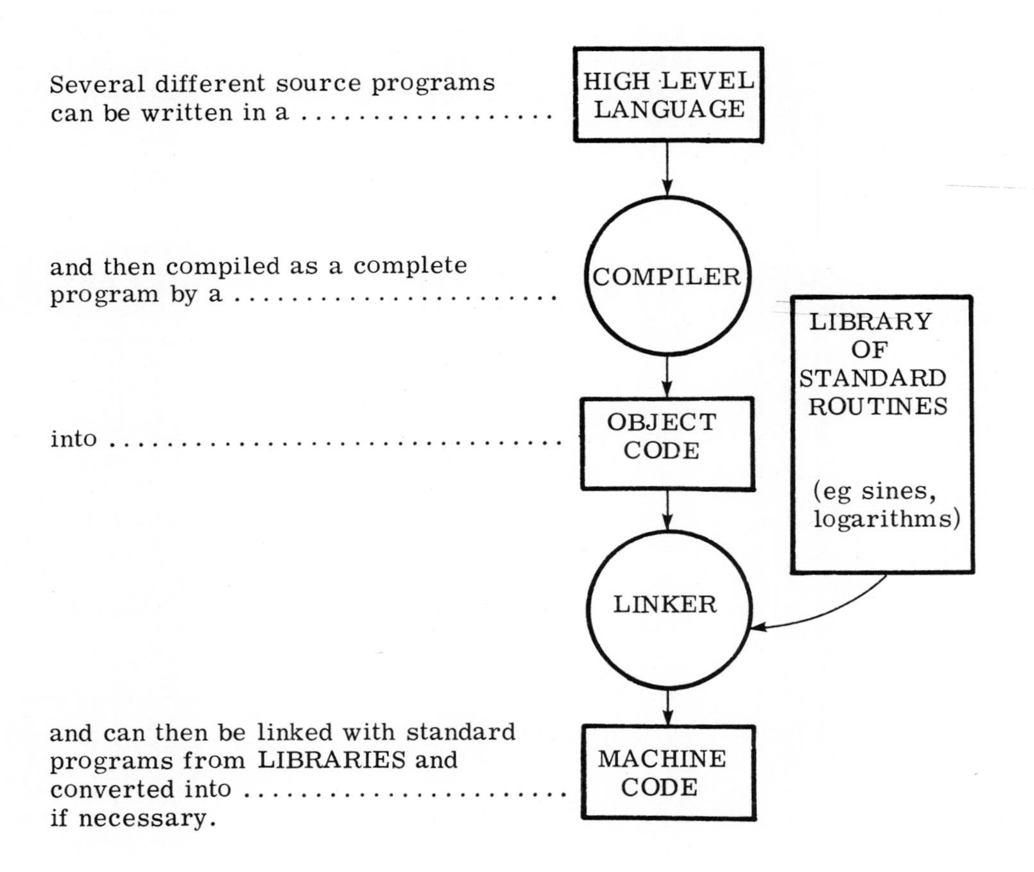

Fig 3.2 The compiler

Programs written in high level source languages can also be rendered into machine code instructions by an INTERPRETER. This examines each high level statement, one at a time, and implements it in machine code as necessary during the live running of the program. An interpreter program is easier to write than a compiler, which is a complex program in comparison and used to create a complete program in machine code and often to introduce, or LINK, other standard sub-routines from program libraries. The interpreter is also much cheaper, but will be necessarily slower to run; it has wide application in microprocessors and in introductory and teaching languages.

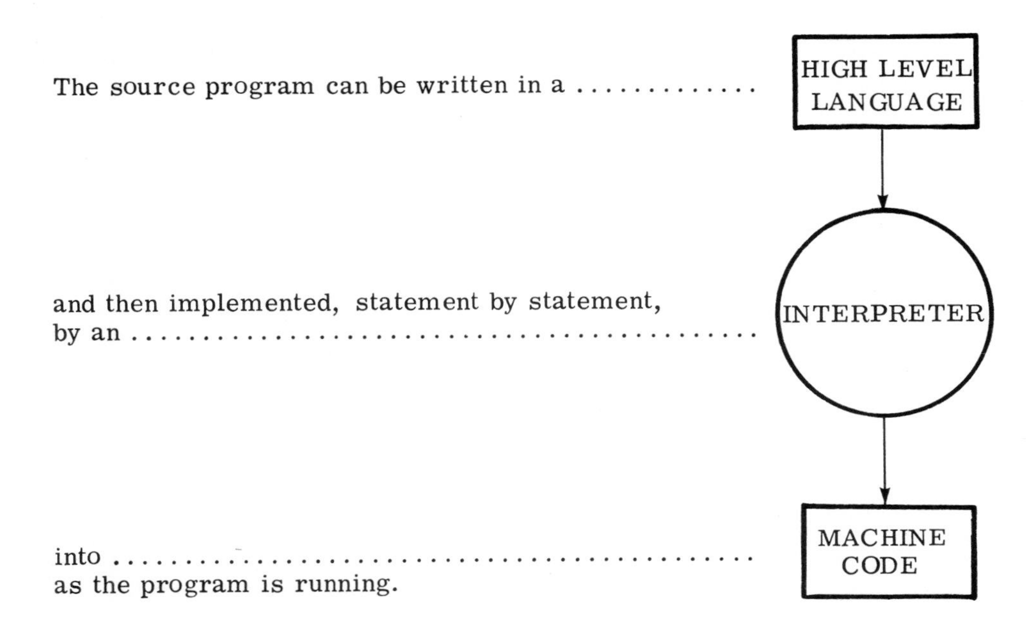

Fig. 3.3 The interpreter

Unfortunately not all users have similar requirements. Some need computers for scientific work, some for commercial or business use, and some for straightforward mathematics; it was thus impossible at first to conceive of one high level language which would prove a panacea to all users. Indeed no single organisation involved in the development of these languages could possibly have even understood the total requirement. The inevitable happened and many high level languages developed, some out of scientific need, others out of the commercial foresight of computer manufacturers. A brief summary of each of the better known high level languages will serve to demonstrate the breadth of development.

ALGOL

The name ALGOL is derived from the initials ALGOrithmic Language: this is a European programming language specifically for mathematical applications. An algorithm is precisely defined as a procedure for solving a problem. ALGOL can be compared loosely with Elliot autocode, but has many fewer restrictions on the use of symbols and operations that can be specified, and its instructions have a general resemblance to algebraic formulae and English phrases.

The structure of an ALGOL program is in the form of a series of consecutive STATEMENTS or INSTRUCTIONS terminated by semi-colons, and a series of statements can be combined to produce a compound statement by enclosing them between the words BEGIN and END.

Every data item processed in an ALGOL program is termed a VARIABLE and is assigned a name or identifier, such as 'X', by the programmer.

Each operation the program is to perform is represented by a statement in one of two ways:

1. An ASSIGNMENT statement, which can have the form

   Variable := Arithmetic Expression

   where ':=' means 'becomes equal to' and the 'Arithmetic Expression' will include variables, numbers, functions and OPERATORS such as +, -, which designate different operations. For example:-

   ```
   X := ( -B + SQRT (B ↑ 2 - 4*A*C) ) / (2*A)
   ```

   This is a common formula; ↑ designates the exponentiation or the raising to a power. The letters A, B, C, and X represent variables and 'SQRT' represents the function provided for calculating square roots. The compiler recognises these symbols and translates them into the appropriate machine code.

2. A CONDITIONAL statement which allows different paths through the program to be followed if specified conditions are satisfied. For example:-

```
IF      X = 3       THEN
BEGIN
        Y:= 4;
        X:= X + 1
END
ELSE
        Y:= X - 1;
```

Groups of similar items of data can be arranged as ARRAYS with a KEY used to identify individual items. The DIMENSION of the array is a measure of the number of keys required to identify an item, for instance an array of the days of the year will be one-dimensional if each day is identified by its number (eg 32 relating to the first of February), but two-dimensional if the day is to be identified by the day and the month, (eg 1, 2).

The structure of ALGOL is of value in that it permits different parts of a large program to be written by different programmers with little risk of confusion and has a potentially economical way of utilising store.

FORTRAN

The name FORTRAN is an acronym for FORmula TRANslation. It was developed in the United States at the same time as ALGOL was being developed in Europe. FORTRAN is a language for scientific work and once again uses a combination of algebraic formulae and English statements of a standard, readable form. A program consists of variables which are allocated alphanumeric names, and statements for execution. Each statement may optionally be preceded by an identifying statement number and can take one of two forms:

1. An assignment statement which takes the form

   Variable = Expression

   Where the expression can be either ARITHMETIC and can include variables, numbers, functions and normal operators, and can form arrays, as in ALGOL, or can be LOGICAL and include all the arithmetic expression facilities but with added operators such as AND, OR, NOT, the logical operators. Using the same formula as was used for ALGOL, an example of an arithmetic assignment is:-

   ```
   47 ROOT = ( -B + SQRT (B**2 - 4*A*C) ) / (2*A)
   ```

   Here the word ROOT and the letters A, B and C represent variables, SQRT the function provided for calculating square roots, * multiplication, and ** exponentiation. '47' is the line or reference number of the instruction.

   An example of a logical assignment statement is:-

   ```
   COST = A.OR.B
   ```

   In this case the variable COST will be given the value true or false according to the logical values of A or B.

2. A CONTROL statement which enables the program to branch to other statements to construct LOOPS. An example of a control statement is:-

   ```
   23 GO TO 48
   ```

   which is unconditional and forces the program to jump to statement 48 of the program for some reason. Alternatively the BRANCH may be used if certain conditions are fulfilled, for example:-

   ```
   36 IF (A.GT.B) GO TO 50
   ```

   which will cause a branch to statement number 50 in the program if variable A is greater than variable B.

There is a good deal of variety and many levels of complexity available in FORTRAN statements, making the language both flexible and popular for scientific applications.

COBOL

COBOL, COmmon Business Oriented Language, was devised to overcome a problem identified by the US Department of Defence, which found itself involved in a massive investment programme involving a large number of different types of computer. It realised, as the number of computers grew, that there would be a requirement for some form of commonality between them and that some standardised method of programming was vital. A language, COBOL, was devised for which all the computer manufacturers involved had to provide compilers to allow programs written in COBOL to be run successfully on their machines.

A COBOL program is written in English of a standard, readable form and in four divisions:

> Identification division, containing descriptive information to identify the program being written.
>
> Environment division, which details the specifications of the computer required to run the program such as the size of memory and number of peripherals needed.
>
> Data division, which allocates identifying names to all the data elements used and defines all the requirements for files and working store.
>
> Procedure division, containing all the instructions to solve the problem.

The instructions consist of RESERVED words which have special meanings to enable the compiler to equate the correct machine code, and the identifying labels for the data specified in the data division or other sections of the program. An example of a COBOL instruction is:

```
ADD ALLOWANCES TO NET-SALARY
```

Here ADD and TO are reserved words instructing the compiler to generate the necessary machine code to perform an addition and ALLOWANCES and NET-SALARY are identifying labels of units of data specified in the data division. Another example might be:

```
IF TOTAL LESS THAN MAXIMUM GO TO SUPER
```

where a branch to another part of the program called SUPER is necessary when the contents of a store called TOTAL are less than the contents of a store called MAXIMUM.

COBOL is fairly easy to learn and programs written in COBOL are relatively easy to test and to understand; they can be run on any machine where a COBOL compiler is available.

COBOL compilers were naturally produced mainly for computers manufactured in the United States. The limitations of COBOL and the forces of national and commercial pride were recognised and for these reasons even better high level languages were sought. Several manufacturers developed their own languages with the aim, presumably, of standardising programming on their range of equipment. Consequently, an owner of one of their machines was forced to buy any replacement from them as a result of natural reluctance to restrain all his programmers in a new language and rewrite all his programs. The problem remained to devise a language sufficiently general for all purposes; one attempt was made in the development of PL/1.

PL/1

PL/1, Programming Language 1, was developed with the intention of incorporating and combining the best features of several other languages including COBOL and ALGOL. The commercially based languages on the one hand are good at handling large amounts of data. They can input and display it in very sophisticated ways, whereas the scientifically oriented languages have enabled highly efficient and sophisticated algorithms to be written. PL/1 attempted to combine the data handling capability of the commercially oriented languages with the problem-solving capability of the scientific languages, to aid the commercial world with its increasing amounts of mathematical analysis, and the scientists with the larger volume of data being manipulated.

PL/1 has three distinct advantages. First that it has a modular structure: a programmer need really learn only the rules concerned with his own particular area of interest; different problems being solved by the use of extensive subsets of the language. Second, that it has a FREE-FORM structure. This means that no special documentation is necessary because the significance of an instruction depends mainly on its format and not on its position in the overall program framework. Third, that a default procedure is incorporated whereby a poorly defined input is given, to a large extent, a likely valid interpretation. PL/1 does, however, have limitations in that it is complex and in total takes a long time to learn; it also requires a very large compiler.

CORAL

CORAL, Computer Oriented Real-time Application Language, was designed initially at the UK Royal Radar Establishment (now the Royal Signals and Radar Establishment) in 1966. The aim was to design a standard language which would permit programs to be run with equally high efficiency in any type of computer and in all applications. CORAL does permit the use of non-standard statements for any part of a program where it may be important to exploit particular hardware facilities; it was considered that in such applications some easing of high level language ideals would be permissable if, in return, there were an effective increase in speed. Such facilities were identified as urgent requirements, particularly in industrial and military areas.

It is a general purpose language for use by skilled programmers and is very suitable for writing compilers as well as for direct application. The main improvements in procedures compared with ALGOL lie in the maximisation of machine code or object code efficiency.

A CORAL program consists of symbols, such as BEGIN, END, 1, 3 etc and arbitrary identifiers which are strings of characters. Identifiers are names referring to 'objects' classified as follows:

data	-	numbers, arrays of numbers, tables
places	-	labels and switches
procedures	-	functions and processes.

Each program can be divided into segments and each segment is in the form of a 'block' or sequence of declarations punctuated by semi-colons and bracketed by BEGIN and END.

BASIC

Beginners' All-Purpose Symbolic Instruction Code (BASIC) is a conversational programming language, developed at Dartmouth College, that uses simple English language statements and familiar mathematical notations to specify operations. The inexperienced user finds it simple to learn but at the same time it provides sufficiently powerful facilities to satisfy the requirement for a wide variety of on-line applications.

A BASIC program is input as a series of lines, one line at a time, and each line is validated by the system. If any errors are found in the format of a line, the programmer is informed immediately by means of an error message in order that it may be corrected. Each line contains a number followed by a sequence of characters: the number is called a 'line number' and the characters are known as a 'statement' as in other languages. The line number specifies the order in which the statements are to be obeyed and also acts as a label to enable control to be passed from one part of the program to another. The statement specifies the operations to be carried out. A separate line is used for each statement and line number.

Each statement starts with an English word indicating the function to be performed, such as PRINT or GO TO, followed usually by the item or items to be operated on. Spaces between groups of letters and digits have no significance (except in character strings such as the number "100"). An example of a simple program is as follows:

```
100 PRINT "INPUT GROSS PAY"
110 INPUT G
120 PRINT "INPUT TAX"
130 INPUT T
140 LET N = G - T
150 PRINT "NET PAY=", N
160 END
```

In the example the task is to read into the computer a man's gross pay and tax and to output the net pay (gross pay - tax). Lines 100 and 120 will cause the phrases 'INPUT GROSS PAY' and 'INPUT TAX' to be displayed to the operator as a prompt. After the first phrase the operator will input a figure for Gross Pay which line 110 instructs the computer to place in a store to be called 'G'. After the second phrase the operator will input a figure for Tax which line 130 instructs the computer to place in a store to be called 'T'. Line 140 instructs the computer to take the contents of 'G' and 'T' and subtract one from the other and place the result in a store to be called 'N'. Line 150 first causes the phrase 'NET PAY=' to be displayed to the operator, followed by a display of the contents of store 'N', the actual value of the Net Pay.

This is a very simple example but it serves to show the very easy style of BASIC. It should be noted that BASIC is mainly available as an interpretive rather than a compiled language and programs therefore take longer to run since each statement is translated afresh into machine code every time the program is executed.

PASCAL

Pascal has increased in popularity as a language since about 1974 and has many desirable features including its ease of use to produce efficient compilers. It is also very popular in the field of systems software and more recently in the programming of microprocessors. Pascal was developed by Professor Wirth in Zurich in about 1970 and was named after the mathematician Pascal to whom reference is made in Chapter 1. Broadly Pascal allows the structure and detail of a program to be expressed in straightforward terms of the data to be processed and the actions to be performed; so problem solving is simpler.

A Pascal program consists of a series of statements or instructions which the CPU will obey step by step. Each statement is written in reasonably clear English and algebraic expressions, however the number of legal statements is quite small and this has given rise to some criticism. The Pascal vocabulary consists of keywords with fixed meanings, such as BEGIN, END, IF (called reserved words), names of objects such as COST, DATE, TIME (called identifiers) and other normal symbols such as + - . The identifiers must be specified and given meanings at the beginning of the program and they must start with a letter but can be followed by a combination of numbers or letters (eg C72B6 or SALARY). Each statement is terminated with a semi-colon. An example is:

```
PROGRAM P962 (INPUT, OUTPUT);
VAR   X, Y, Z : INTEGER;
BEGIN
      READ (X, Y);
      Z := X * Y;
      WRITELN (Z)
END.
```

This is the Pascal program to solve the problem of multiplying two numbers together and printing out the answer. In the example:

Line 1 contains the name of the program (P962) and a bracketed set of names concerned with data and results.

Line 2 indicates that identifiers X, Y, Z have been given to three variables or stores; in this example the identifier in each case is a single letter. The statement also indicates that the values to be given to each variable (ie the number to be put in each store) will be an 'integer' or whole number value as opposed to 'real' where the variable may take a value with a fractional part, for example 47.26.

Line 3 indicates that the program follows.

Line 4 specifies that two successive values are to be taken from the data input and assigned to the two variables X and Y.

Line 5 assigns the value produced by multiplying the values of variables X and Y (ie the contents of stores X and Y) together, to variable Z, ':=' indicating the assignment 'becomes equal to'.

Line 6 specifies that the value assigned to each variable inside the bracket is to be printed out on the output of the computer. (In this case Z only).

Line 7 is necessary to indicate completion of the program.

From this simple example it can be seen that a Pascal program is easily understood. For this reason and the fact that it is easily maintained, Pascal has been selected by several manufacturers and institutions for experimentation in an attempt to increase its application across wider fields and in particular across particular product ranges. A notable example is the development of Pascal at Texas Instruments Inc over several years after it was selected as their standard corporate systems language. As Pascal's popularity increased it tended to be oversold as an ideology rather than as a programming language. Such promotion has quite naturally provoked criticism from those who believe in "horses for courses": in other words they would rather see a variety of languages, with preference depending upon particular circumstances and programming tasks.

Problem Areas

The disadvantages of all high level languages lie in the relatively inefficient use of store when compared with programs written in machine code. Also there is a lack of flexibility caused by the restrictions on the number of instructions available in a standardised language.

The generality of use of higher level languages still depends very much on the ability of the computer manufacturers to develop efficient compilers because each different machine normally needs its own compiler. For this reason various dialects of each high level language have evolved! The reason is that there is difficulty in producing compilers for some machines to convert the high level language into the appropriate code and so take advantage of the different capabilities available in machines.

It is possible that the ideal general language will never be realised, for how can the opposing aims of universal understanding and perfectly matching compilers for all machines be rationalised? An attempt is being made to develop a language to a well defined standard in the defence area. The language is called Ada and is referred to in Chapter 8.

A PROGRAM'S DEVELOPMENT CYCLE

Introduction

It may be useful at this stage to try to knit together various computing concepts, such as programming languages, hardware, compilers and operating systems by describing briefly the sequence of events through which a typical fairly simple program passes from its initial conception to its operational running.

Phases

The main steps in a program's development cycle are as follows:

- problem definition
- problem analysis
- program design
- specifying test data
- writing the source program
- preparing source program and data
- compilation
- linking and loading
- object program running
- finalising documentation

These steps will be briefly described below. It should be said at this stage that very few programs go straight through all these steps without continuous iteration or looping back. For example the first compilation of a program will normally produce errors which must be corrected: as a result the source program needs to be changed and then the program to be recompiled. The flow diagram in Fig. 3.4 shows this clearly.

Problem Definition, Analysis and Program Design

Problem definition, problem analysis, and program design are normally grouped together under the general heading of Systems Analysis and will be discussed in Chapter 4. Depending on the environment and the type of program they could all be carried out by one person, who could also write the program, or by a number

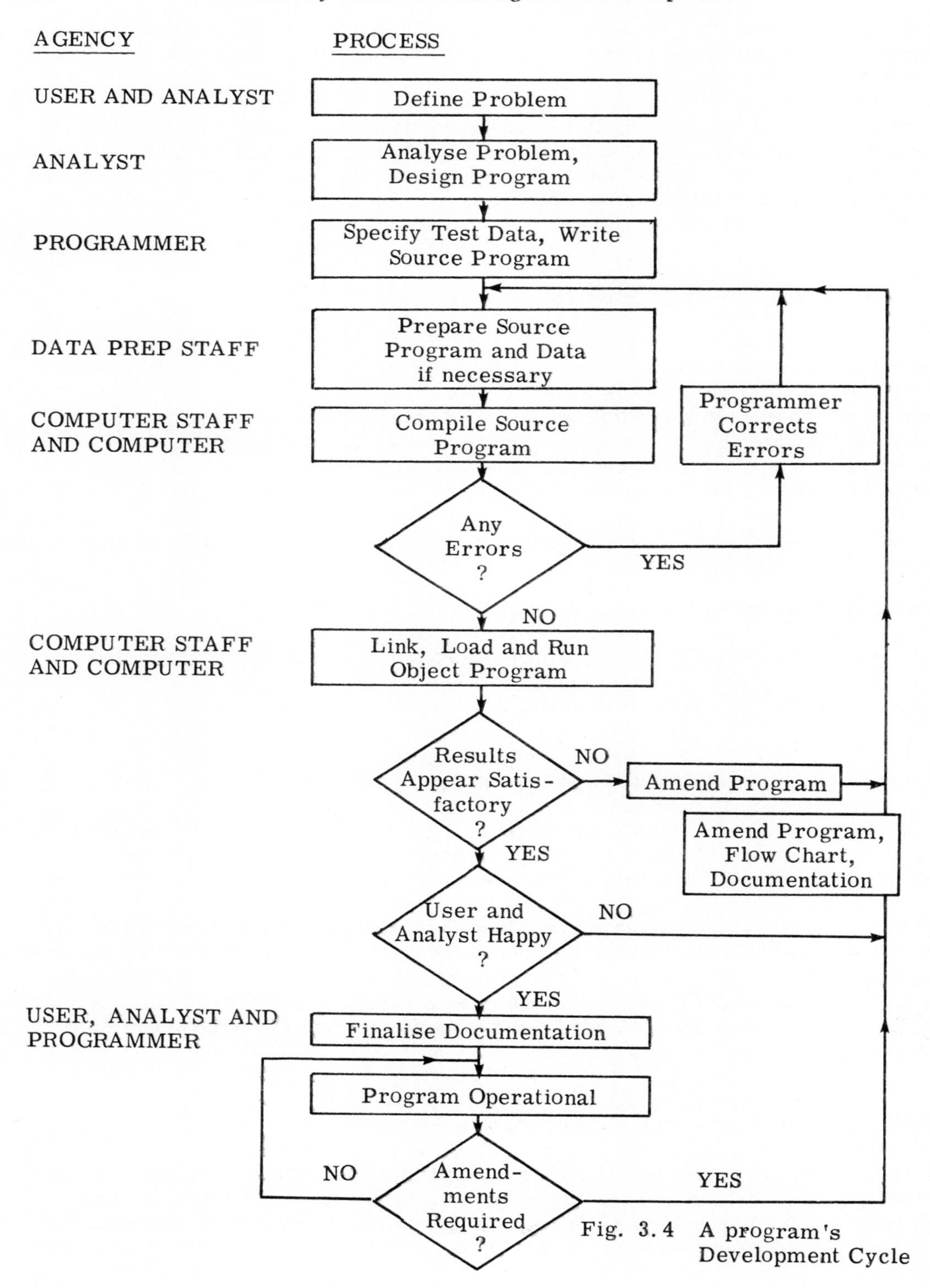

Fig. 3.4 A program's Development Cycle

of people. In any event the definition of the problem is the responsibility of the user. If the user is not the person who will carry out the analysis, then the problem definition should be done in conjunction with the analyst, who is a trained specialist. Problem analysis and program design involve careful consideration of the best approach to putting the problem on the computer and then the design of the processing steps and the data structure which the program will incorporate. The output from the program design should be a flow chart setting out the steps in processing, and some meticulously detailed documentation describing the data structure.

Specifying Test Data

Before the program has been written, test data should be prepared to test it under all foreseeable conditions. This, as may be readily imagined, is by no means easy, particularly as the program may have many conditional statements. However it is critical if the end product is to meet the stated requirement. The discipline of testing software in modules will indicate errors which might otherwise take much time and trouble to locate subsequently.

Writing the Source Program

Today almost all programs are written in the high level programming languages discussed earlier such as COBOL, ALGOL, FORTRAN, CORAL etc. The general advantages of writing a program in a programming language as compared with writing in the machine's own binary code are, first, that it is very much easier to write, read and check and, second, that if any errors are made they are much easier to detect and correct.

Preparing the Source Program and Data

Most programs written today will be input directly to a computer, instruction by instruction, by a programmer seated at a VDU. In a card or tape fed system, the source program and data in manuscript form must be transcribed into a machine readable form on the card or tape.

Compilation

Before the source program can run and produce results the translated version of the program in the computer's own binary code, or object program, must be put into the store. This is achieved in two stages; they are compilation followed by linking and loading.

To start the first stage, the compiler program must be brought out of its own store. When the compiler has been loaded, another message is passed to start it running. Now, as far as the compiler is concerned, the source program is merely input data to be processed. This process has three main outputs:-

a compiled version of the source program
which is now held in store

a listing of the source program

a listing of error messages

If there are errors, the process stops here and the listings are available to the programmer for correction of the source program and recycling.

Linking and Loading

The compiler program has now carried out its main task, its last actions being to trigger its own deletion and to call up another program which has the functions of incorporating the routines invoked by the source program, such as a standard printing routine, which are not translated at compile times but are held, already in binary code, in a special library store. This process is known as LINKING; the consolidated object program is now loaded into store and the executive informs the programmer that the object program is loaded (LOADING).

Object Program Running

The test input data for the object program is now loaded into the computer and the programmer orders the start of the object program. The object program reads the data, produces results and continues running until either an error occurs, when it can be stopped, or until completion. Results are then returned to the programmer who checks them. If they are not correct the source program is amended and recycled. As soon as the program performs correctly the analyst and user are consulted to establish that it meets their requirements. Any corrections as a result of this consultation are carried out and the program is now 'operational', although of course further amendments may well be made during operational running.

Documentation

The final step, far too often lightly dismissed, is to finalise the program documentation. This should include at least a description of the problem, flow chart, data specification, operating instructions and program notes.

SYSTEM SOFTWARE

Introduction

System software can be considered as those programs and procedures designed to ensure that the system, including its peripherals, compilers, executives and all its store, function effectively. It will therefore be concerned with the efficient allocation of time and the system's resources between all the contending applications programs.

Direct Memory Access Controller (DMAC)

Transferring of data from backing store to main store can be achieved without involving the CPU by employing a processing function known as a DIRECT MEMORY ACCESS CONTROLLER (DMAC). A microprocessor can be used in this role.

Multiprogramming

The ability to run programs apparently simultaneously was mentioned in Chapter 1. An EXECUTIVE program or OPERATING system is used to control peripheral sharing and the operations concerned with loading and communicating with programs. When a program requires that data is transferred to or from a peripheral, it obeys an instruction which passes control to the executive program; the executive initiates the transfer and the original program is suspended while the transfer takes place. The executive consults the order of priorities allotted to the various remaining programs present in the computer and allows the program of highest priority to proceed to make use of processor time which would otherwise be wasted. When the transfer relating to the first program is complete an automatic interrupt occurs and the executive decides whether or not, depending on priorities, to suspend the currently running program and continue the first program.

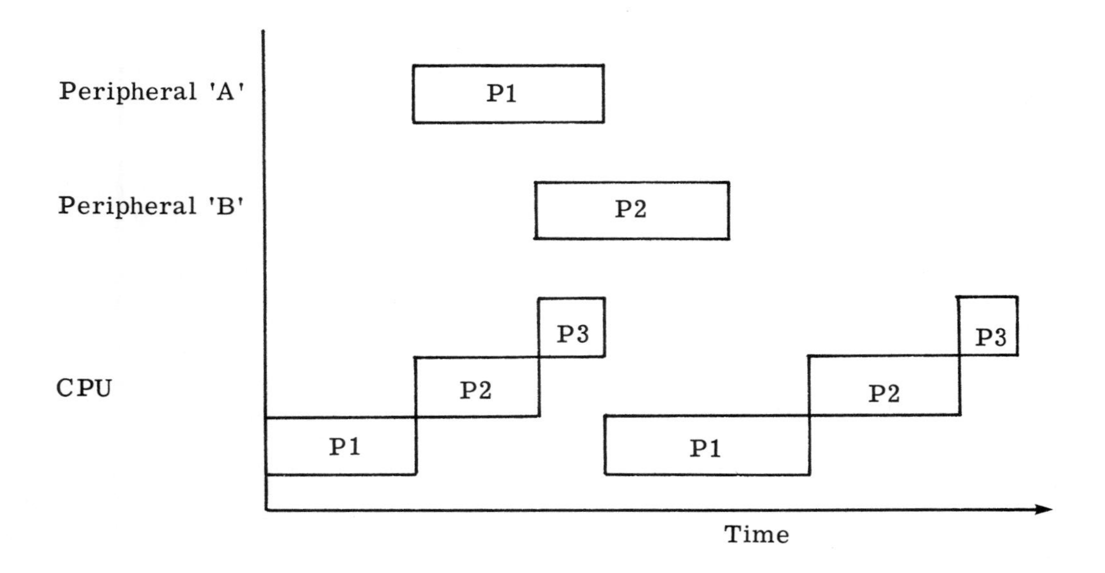

Fig. 3.5 Multiprogramming

Figure 3. 5 shows a simplified graph of peripheral usage against time. In the diagram program P1 went into the CPU but after a while needed to use peripheral 'A'; the executive instructed 'A' to start operating and while P1 was then waiting, P2 was introduced into the CPU by the executive. P2 then needed to use peripheral 'B' . . . . and so on. Any of the three programs P1, P2 or P3 could have been assigned a priority, but in the example priority is assumed to be in ascending order and thus, once P1 had finished with a 'A', it interrupted P3 in the CPU.

Databases

In order to manage such a system, which has parts of various applications apparently moving around the system simultaneously, it is obviously necessary to have a very well ordered and structured method of holding and indexing all those programs and all the associated data. The normal approach is for all users and applications programs to use one common set of data, known as a DATABASE. A database consists of files of data so structured that each applications program may go to a file, draw data from it, work on the data and update it. However it is important that the applications programs must not constrain the design of the file or the contents of the file because, by definition, both the files and data are for general use.

Real-Time Systems

An ON-LINE system is one in which the input data enters the computer directly from its point of origin and/or the output data is transmitted directly to where it is to be used. On-line systems have been made possible by advances in communications techniques and are now very common; there is virtually no limit to the distance possible between the user and the CPU. Typical applications are the linking by telephone lines of all major branches of a bank to a central computer at a head office or the linking of computers in battlefield command and control systems by trunk or net radio.

A REAL-TIME system is one which controls an environment by receiving data, processing it, and returning the results quickly enough to affect the functioning of the environment at that time. It is speed that is vital in a real-time system, hence most real-time systems are on-line, which means that they are directly linked to the CPU to eliminate data preparation time. However on-line systems are not necessarily real-time. For example, a bank clerk may use the computer terminal in his branch to prepare a monthly payroll; there is probably no time constraint and such work is called REMOTE JOB ENTRY. On the other hand the same clerk may use the terminal to enquire, on behalf of a waiting customer, if there are sufficient funds in his account to cover a high value cheque. The latter is an on-line, real-time application. Real-time systems are good examples of the requirement for a multiprogramming capability; systems software in a real-time system has to control a large variety of programs in different states of execution and of varying priorities.

Weapon systems and command and control information systems are all examples of real-time systems. In these, data to be processed is captured by devices of different characteristics and subjected to widely differing processes and applications software. However their control, or systems, software has much in common. The common tasks which can be identified are:

Activation and suspension of the programs under control.
Supervision of the interfacing between programs of varying priority.
Management of queues for system resources or peripherals' input/output control.
Management of large, structured, sets of files.
Self monitoring of own performance.

Discussion of such a piece of executive software will inevitably lead to dividing it into modules or sections and it is in this form that it is constructed, as is the software it controls. Furthermore, associated with all the modules will be a priority.

A vital characteristic of a real-time system is reliability and this feature must be designed in from concept. Reliability of a total system is measured in terms of the mean time between failures (MTBF) or, more sensibly, in terms of system availability which will include not only MTBF but also the time taken in repair and recovery of the system (MTTR - Mean Time to Repair). Hardware duplication has often been used to ensure high availability and the redundant hardware used for other tasks until it is required to be switched in as a result of primary hardware failure, this concept, however, is often not viable in military operational equipment and a more common fall back position will be the provision of spare capacity and power in each cell or module of a system to enable it to take over the role of a neighbouring cell, should it fail. Fall back procedures must also be designed into individual cells in order that some service, albeit downgraded, can be provided. In a multiprocessor configuration this can be achieved by giving added tasks to individual processors should another processor fail. In addition, the user may feel that manual back-up procedures are necessary; to aid this frequent 'dumps' or printouts of the contents of the database are taken to enable the human operator to continue. In many cases the back-up systems are proved to be unnecessary and there is a strong lobby of informed opinion that believes in 'living with the technology that has been developed' or rather 'having faith'.

The development of real-time processing has produced new concepts and makes heavy demands on software and hardware. The challenge lies in designing a system that can accept stochastic or random inputs, process the data and present the solution in a very fast response time. The added need for exceptional reliability, especially in military systems, merely makes the challenge that much greater.

Debugging

Debugging is the composite technique of detecting, diagnosing and subsequently correcting errors, known as BUGS, which occur in both software programs and hardware. Software errors will be of one of two types; they will be either syntax or logic errors, for computers work only on logical statements which are presented to them in a correct syntactical form. An error in logic is caused by an incorrect understanding of the requirement, for instance an incorrect setting down of a formula. An error in syntax is caused by the incorrect coding of a program and can be as simple as the mis-spelling of an instruction. A logic error will be executed by the computer but the result will be wrong whereas the computer will reject a program containing a syntax error and the program will not run.

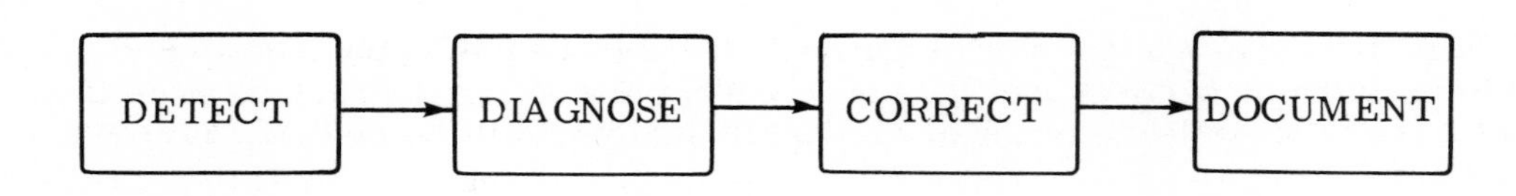

Fig. 3.6 Debugging

Logic errors are detected by programs failing to produce expected results from sets of known test input data. Test data is input to the computer directly or in the form expected in normal running and will be as comprehensive as possible. Syntax errors will be observed when the high level language is compiled into machine code; compilers will reject such offending errors and print some indication of the type and location of the error.

Programmers frequently diagnose logic errors by re-enacting each stage of the flowchart and pretending to be the computer. They can also use some sort of software diagnostic routine which will perform the program in the normal fashion on the computer but will record the action taken by each instruction; there are many versions and complexities of such diagnostics. Syntax errors are normally, as already mentioned, identified by the computer.

The program may then be corrected by removing incorrect statements and substituting correct ones or by application of a PATCH which involves removing incorrect statements in the object code and replacing them with correct object code statements. It is essential, of course, that documentation should reflect all such amendments and corrections.

File Processing

File processing is a collective term which covers all the operations associated with the creation and use of files or sets of records. The operations are:

1. CREATION, which consists of the initial collection of raw data and the organisation of this data into files.

2. VALIDATION, which is the process of checking to ensure that the data in the files falls within the prescribed limits, for instance only numeric data may be entered in a prescribed numeric only file.

3. COMPARISON, in which the data held in two or more files is compared as a check for similarity or discrepancy.

4. SORTING, which involves the ordering and presentation of data from files which is to be processed in a predetermined sequence. Sorting depends upon the sequencing of the data items against a key. An example is the alphabetical listing of names in a telephone directory; the key is the surname.

5. MERGING or COLLATING, by which a single file is created from two or more files which already exist and have been arranged in the same sequence by sorting on a common key.

Utility Programs

File processing is carried out by UTILITY programs which are provided to perform the operations on the files of data; they treat the files as units and are not involved with specific items of data. They have three main roles: first the transferring of data; second the copying of data; third the re-sequencing of files.

The transferrence of data involves the transferring of files of data from one storage medium to another. For example data can be transferred from paper tape to magnetic tape, normally as part of a chain of events to put data required for processing into faster storage. It can also be transferred from magnetic tape to the store associated with a printer, where it may be held until required to be printed out.

Copying data may be necessary as part of a security plan for a particular system in which it is essential that files to maintain a record are periodically copied in case of failure of the normal medium. The re-sequencing of files covers the mechanisms for placing files in any required sequence and the ability to manipulate data which is not held in a serial sequence. Other utility programs can be designed to check the performance of stores and to validate the files.

APPLICATIONS SOFTWARE

Applications software comprises those programs closely conforming to the user's job. The user has a great deal of influence in the development of them, as opposed to the general purpose systems software which manages the operation and working of the total system. It is possible to divorce systems and applications software almost completely during development providing that the interface requirements are carefully defined at a very early stage; a complex system may well have many application programs. Each application package module can be written by a different programmer or team of programmers independently but within the interface rules. It should be noted, however, that in very small single purpose computer systems there may be a requirement for only a single item of applications software and no systems software at all.

SOFTWARE MANAGEMENT AND PRODUCTION

"It is common knowledge that software development projects rarely meet cost-benefits originally projected, usually cost more than expected, and are usually late. In addition, the software delivered seldom meets user requirements, often times is not usable, or requires extensive rework. . . . "

A Software Acquisition and Development Working Group (SADWG), July 1980.

A report from the group making this statement goes on to say that the problems are not, however, insurmountable and the key to their solution is in "better management at all levels. . . . "

During the 1960s the term 'software crisis' first became common, due to advances which had taken place in hardware that had not been matched by corresponding advances in the techniques for constructing software. Ambitious systems were delivered years late, performed unsatisfactorily, and proved immensely difficult to maintain and enhance. Tales of failure abounded.

Structured Programming

"As mathematics passed the year 1800 and entered the recent period, there was a steady trend towards increasing abstractness and generality. . . . Abstractness and generality, directed to the creating of universal methods and inclusive theories, became the order of the day".

E T Bell, Development of Mathematics

With the realisation that further hardware developments would demand even more complex software the problem was getting worse and it was realised that no improvement could be expected until a more formal and disciplined approach to software writing was established. Programmers could not be allowed to continue to indulge in their own particular style because under such circumstances it was impossible to organise teams of programmers to build efficient large scale systems. The problem was indeed one of size and complexity. If the required program was within the capability of a particular programmer he could be relied upon to perform a satisfactory job provided he was available to carry out maintenance himself. But as soon as more than one programmer became necessary problems arose, no matter how good the documentation.

Many solutions for dealing with the software crisis have been proposed over the years. Each has had its own contribution to make and, while no panacea has been found to date, enough progress has been made to avert many of the potential catastrophes. Two most important concepts which have evolved are, first, a modular approach to design, which involves breaking down the task which is to be performed into a hierarchy of sub-tasks. To each sub-task is assigned a module of code which has a well defined interface with the remainder of the hierarchy. This approach has a number of advantages. By making the sequence of the program, which is called the MACROSTRUCTURE, mirror more nearly the structure of the task some classes of errors are minimised and the program becomes more readable. Because it is more readable, errors are easier to locate and correct; maintenance and modification is more easily carried out by the author or by someone else. Once the modular design is complete individual modules can be coded by different programmers with a reasonable hope that the whole will ultimately work successfully. Furthermore the modules can be tested separately and in small groups.

Modular design for programming is possible in many language systems which allow independently compilable units. Its use in assembly code programs requires rather more careful planning than is needed for those written in high level languages. It is interesting to notice that FORTRAN, the earliest and most primitive of languages with a claim to be called high level, is especially well adapted to the modular approach. Modularity in FORTRAN is inherent and the programmer has a positive encouragement to write in this way, but this is not to say that the encouragement is always welcome or heeded.

The second concept is that of structured programming, which is an attempt to lay down a set of guidelines for the writing of clean, gimmick-free programs, using modular design and high level languages as far as possible. It has two basic interrelated and parallel approaches. The first approach is learning to program, as opposed to learning a programming language; this involves acquiring a set of techniques for general purpose problem solving. A recommended method is called TOP-DOWN ANALYSIS or STEPWISE REFINEMENT. In broad terms a completely general statement of the task which the program has to perform is made. The next stage is to identify a set of sub-tasks which, carried out in sequence, make up the main task. The sub-tasks are defined by statements in English, written in terms of what is to be done rather than how it is to be done. In performing this process great care has to be taken not to introduce too much detail too soon; indeed there can be levels at which the precise nature of the data structures is immaterial. The stepwise refinement continues until there is sufficient detail to write a representation of the problem in a programming language. A program designed in this way will inevitably be modular. Certain languages, for example ALGOL, CORAL, PL/1, Pascal and Ada are designed to make structured programming easy, the remainder do not. The second approach is to build up a correct program initially by systematic design, and not by endless iterative testing.

Sometimes a BOTTOM-UP approach is used. This implies starting with single modules and often the programmer will not know how the whole system will build up. He can only hope that his program will fit the overall plan.

MASCOT

'Modular Approach to Software Construction, Operation and Test' (MASCOT) is an example of a technique for software production and management. MASCOT is a wide-ranging philosphy which can be applied through all stages of the software life-cycle from design onwards. It defines a formal graphical notation, independent of both computer configuration and programming language, in which the overall design can be expressed. The resultant 'blueprint' provides a good representation of the design which contributes usefully to the process of clarifying the requirement and establishes a clear and unambiguous basis for the remaining stages of the project. This disciplined approach to design leads to a highly modular software structure in which a close correspondence exists between the functional design elements and the constructional elements used in system integration.

There is an orderly program development strategy based on the test and verification of single modules and larger collections of functionally related modules. A small and easily implemented executive program for the dynamic control of program execution provides a standard software environment in which the software can operate. All MASCOT modules, during individual testing, integration testing and in the final operational system are executed in this same environment. This arrangement helps to ensure the validity of the tests and lends itself extremely well to system development through a prototype.

MASCOT is a design method supported by a programming system. It is neither a language nor an operating system although it includes elements that are related to both these aspects of programming. It brings together a co-ordinated set of tools for dealing with the design, the construction, operation and testing of software. It is applicable to a wide range of application areas but is aimed particularly at real-time applications where the software is complex and highly interactive.

We have seen that a principle of structured programming is that overall program design should be performed by a process of top-down analysis. It is essential, too, that the detailed expression of this design in a high level programming language should accord with the principles of structured programming. MASCOT in no sense conflicts with this advice, indeed it supplements it. The basis of structured programming is to make the written program correspond as closely as possible, in its lay-out on paper, with its order of execution by the machine. All programs are reduced to a sequence of operations in which each individual operation is a sequence in itself, thus the emphasis is on sequence and the method is applicable to sequential programs.

Unfortunately complex real-time systems cannot be reduced to simple sequential programs. Rather they consist of a set of parallel, co-operating processes. In other words there are a potentially large number of sequential programs which communicate with each other and with the outside world in an asynchronous manner. Structured programming can, and should, be applied to each of the sequential processes but in order to impose discipline and order on the system as a whole an extra tool is required. A way of achieving a static view of an entity consisting of a succession of executed instructions which appears alarmingly dynamic and unpredictable is necessary, and MASCOT offers such a tool.

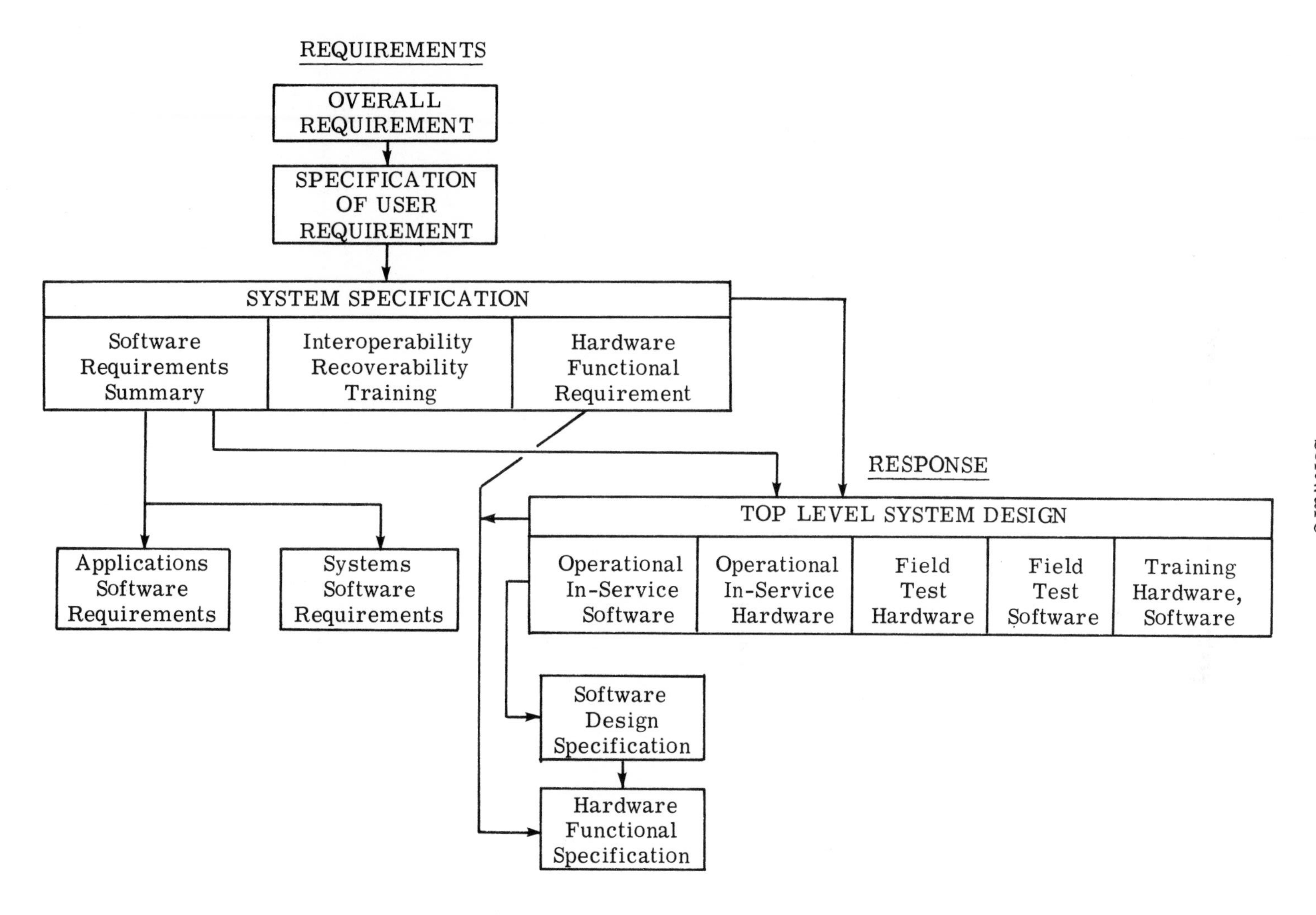

Fig. 3.7 A typical documentation structure

Documentation

It is easy to see that in the development of an ADP project there will be masses of associated documentation in various levels of detail and produced for different purposes. Control of the documentation at all levels is vital to the efficient management of the project and indeed to the eventual successful implementation of the finished product. Methods vary but a documentation structure skeleton could be as shown in Fig. 3.7 where the documentation hierarchy is divided clearly into levels and into 'requirements' and 'responses'.

Development Methodology

On large software projects it is difficult to contain the information explosion that occurs. This information includes the project plan with its allocation of resources including people, details of the design and overall systems architecture of the components to be produced, work schedules, progress reports and in general the multiplicity of inter-related documents.

A lesser explosion can be easily handled on small to medium projects which would take up to, say, 20 man years by conventional manual methods, supported by use of a Project Evaluation and Review Technique (PERT) package. On large projects however, experience shows that the inter-relationships between the components of a system and the activities of the people involved in its development become so many and complex that conventional methods break down. This has been recognised as one of the major causes of cost escalation and failure to complete the task in the planned time.

Information about these inter-relationships is necessary for successful project management and yet this information is so complex that only a computer system is capable of handling it. A database with associated programs for using it is required to record and report the progress of the project, and its many highly inter-related parts, at any one time.

The approach must be to record and monitor detailed day-to-day activity at a much lower level than previously possible and thereby provide management with more accurate information about progress from detail in reports. Individuals and teams do not in practice address themselves to their tasks in a sequence of separate, non-interacting activities. It must be possible to describe highly parallel situations where several project phases overlap, and allow for detailed specification and scheduling of activities and resources.

The database could contain an information model that represents the design and is regularly consulted by the system designers who enhance and amend this model. In addition, the database could offer the ability to record a system architecture which may be used either with a structured programming methodology or for the validation and structuring of a design as it progresses.

The information model would be a record of these design and construction decisions together with cross-references to all the relevant documents which are themselves contained and itemised in an easily retrieved way. As well as a project control aid, it is a cost-effective tool for maintaining and enhancing a software or other complex project over a long period.

Commitment and well-placed effort are required to guide a complex project through all its stages towards a successful in-service life. Tools for project managers and system designers to integrate the management functions with the detailed work of the project are absolutely essential.

Such software aids are becoming available. For instance the SDS Software Development System conceived by RSRE and Software Sciences Ltd (UK), and first implemented at RSRE, which is being used on BATES, WAVELL and PTARMIGAN, three major UK military projects. These aids can provide assistance in the following areas:-

1. Project management. Facilities for both strategic and day-to-day planning are needed on as many levels as required to reflect the management hierarchy. Scheduling of the project plan can be achieved by a multi-level Critical Path Analysis technique. This can be used to supplement or replace conventional PERT charts. Cost control and invoice production can be obtained by the analysis of data on progress and expenses, in conjunction with resource costs.

2. Requirements analysis, design and coding. There are methods for recording and verifying design decisions and assumptions, evaluating their consequences and ensuring their consistent implementation. Recording, integration and reporting on the changing status of all system components in relation to the whole provides project leaders with complete objective management information, and designers and programmers with up-to-date centralised specifications.

3. Configuration and change control. This provides for the recording of the composition of various system releases in terms of the versions of their components. Furthermore, the change requests can be cross-referenced to the relevant versions of those components.

4. Maintenance. A system is required to provide access to complete, fully cross-referenced records of system structure, specification, and documentation. This is to facilitate maintenance of the system throughout its life.

5. Documentation. A facility is required to structure descriptions and other data related to components and activities into standardised printed documentation; further, it needs to cross-refer from project documents, components and activities to external documents such as statements of user requirements.

6. Monitoring and reporting. This is the provision of a comprehensive range of input and output facilities to ensure that the database accurately and completely reflects the project requirement. It also allows effective access to the information it contains, possibly incorporating prompting and graphic facilities.

SOFTWARE QUALITY ASSURANCE

The expense and reliability of software has been one of the largest problems which has faced the development of computers. Indeed the expense in many cases has been a symptom of the unreliability. The problem is not new, for even in the early 1950s it was recognised that much of the time spent in software development was taken up in identifying and correcting errors; this situation can of course become critical when the errors escape detection for a long time: building on an error can be very expensive.

It is important, in defence of the software writers, to examine the definition of an 'error'. For instance there are tales of an early warning system which mistook the rising moon for a missile heading across the northern hemisphere; the user was apparently quick to condemn the software. However it could be that the specification stated very precisely that a warning must be given of any object appearing over the horizon and not readily identified as a friendly aircraft! There is therefore some doubt as to the presence of an 'error' in the software.

There is a basic situation which must be understood when considering the presence of errors in software: it is that if software fails to perform according to its specification then there must be an error in it. However, if software does perform according to its specification it does not follow that the complete system is error free. Imagine an air defence system built to handle up to 50 targets faced one day with 55 targets; the software cannot be blamed if the system collapses.

What then is software reliability? It is probably defined fairly as the probability that software will perform for a particular period of time without failure, weighted by the cost to the user of each failure encountered. This means that reliability is a function of the impact of failure on the user and not merely the frequency or size of the failure. Computers used to control spaceflights have been shown to contain software errors even after incredible expenditure and care. The user must come to terms with the chance of such errors.

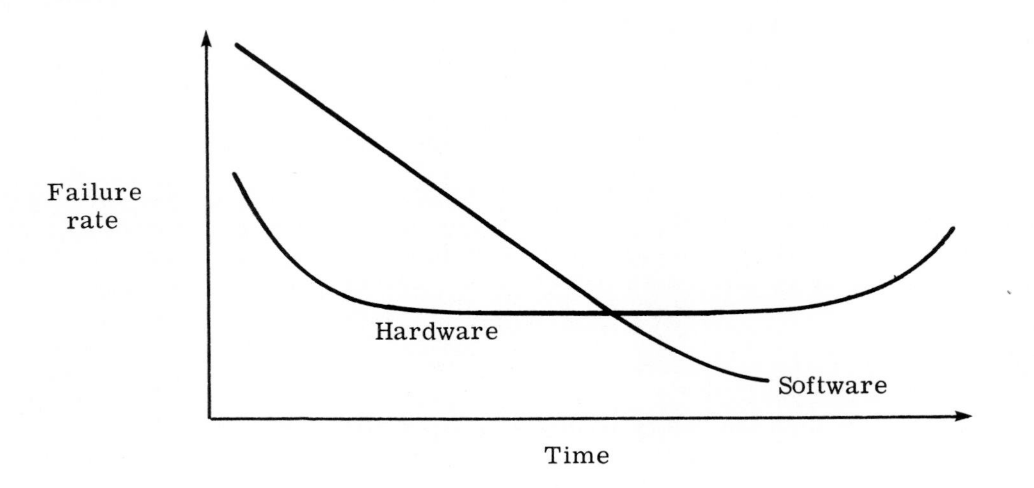

Fig. 3. 8 Comparison of hardware and software reliability

It is interesting to compare the failure rates of software and hardware. Their difference is shown in Fig. 3.8. The hardware failure rate could be that of any electronics component where assembly and manufacturing errors are overcome fairly quickly, followed by a steady but random failure of components until eventually major items wear out and failure rates increase. In comparison, software does not wear out and as each error is detected and corrected so the failure rate falls, if no new errors are introduced either by the correction procedure or by software enhancements added later.

Evolution of Software Quality Assurance

During the 1960s most software advances were in the understanding of compilers, executives and operating systems, that is the functions of the programs being written. High level languages were in their infancy and many programs were written in machine or binary code and it is fair to say that to make a complex program work at all was quite an achievement. In the 1970s more emphasis was placed on the methods used to produce the software such as structured programming. High level programming languages developed and it is clear that in general the higher the level of language the fewer the errors that occur, because a programmer using machine code spends a good deal of time feeling his way through machine details rather than in careful programming. High level languages tend to eliminate many levels of software error by camouflaging most of the idiosyncracies of the particular machine and by the use of fewer statements for particular functions.

Programs have become far more understandable, far easier to change or amend and are in most cases relatively self-documenting. Programs in high level languages are also cheaper because a high level language statement is functionally superior to a machine language statement. Most important of all is the new capability to express and manipulate highly complex data and store structures.

The efficiency of a program is probably a measure of the programmer's ability to select the correct algorithms and data structures rather than his ability to select the right language. There is also no concept of any tolerance in the accuracy of a program, unlike any other engineering design. Structured programming and design have become the order of the day in order to anticipate and isolate errors and to enable programs to be written more cheaply.

These then, are technical advances which have steadily improved the quality of software. There is however the intangible element of the very psychology of the programmer to consider. Software writing is a manpower intensive activity and as such must be affected by the state of mind of the individual concerned. There is a psychological principle which guides an individual when his self image is threatened and in many cases, including that of the programmer, it may manifest itself as an infinite capability never to admit to an error. The answer to this may be to promote an attitude of openness and constructive criticism within a software team. It is important to develop, in parallel, a healthy attitude concerning accuracy and precision; the more errors which are detected early the better. In this area fault detection at module level has proved exceptionally successful but is not without its problems.

Software Management

As the computer industry evolved in the 1970s it became steadily more apparent that efficient management of software development was critical to avoid cost and time overrun and in fact to achieve a viable end product. Analysts and designers started to have more discourse with the user at an early stage and documentation became a way of life in order to provide some ordered structure to a very volatile work area. At the same time software objectives were set in order to provide targets against which progress could be measured: these objectives were concerned with both the software product and the project as a whole and contained, for example:

Cost targets
Test objectives
Reliability levels
Definitions
Project schedules
Compatibility tests
Configuration controls
Security requirements.

Never before had the management of such projects been so clearly defined in this very necessary fashion. Having established the objectives and milestones it was possible to arrange systematic design reviews where progress and problems were reported and the user was able to see what was happening.

The first step in the review of a design to check its quality is for an independent observer to try to identify weaknesses in the design. It can not be done on the spur of the moment but must be thorough and take a fairly lengthy period. The second step is to hold a dynamic review or 'walkthrough'; here individual programs are actually run using test inputs and the flow of data is followed mentally through all the modules. These two steps may appear simplistic but experience shows them to be of immense value.

Software Testing

Testing of software can account for up to, say, 40% of software development costs! (See Fig. 3.9). This demonstrates the important role it fulfils and possibly shows the need for even better software tools and management. One thing is clear, it is absolutely imperative that project plans include detailed testing plans at a very early stage in order to save expense.

The test must be a function of the requirement; in other words designers should be encouraged to 'design to test'. The purpose of testing software is to discover as many errors as soon as possible rather than to catch out an individual programmer. The task has an almost infinite number of paths it could take and so cost effectiveness must be considered; it is not possible to test every possible combination of paths. There are thus a number of guidelines now firmly established to assist testing agencies:

1. Define the criteria for success. All those involved should be aware of the criteria against which success or failure will be assessed. Requirements must contain performance criteria.

2. Be objective. Only impartial examiners should be used. They may have different views on acceptability.

3. Limit the scope. It would not be cost effective to test all cases; a sensible, pertinent selection of tests is required.

4. Document the results. This may be obvious, but unless it is carried out faithfully it will be impossible to repeat the test should the software be amended subsequently.

5. Vary the test data. Tests should not be carried out only on 'expected' data. Efforts must be made to try and outwit the software by feeding it invalid data to see that it can respond. In military systems in particular, robustness of software is vital.

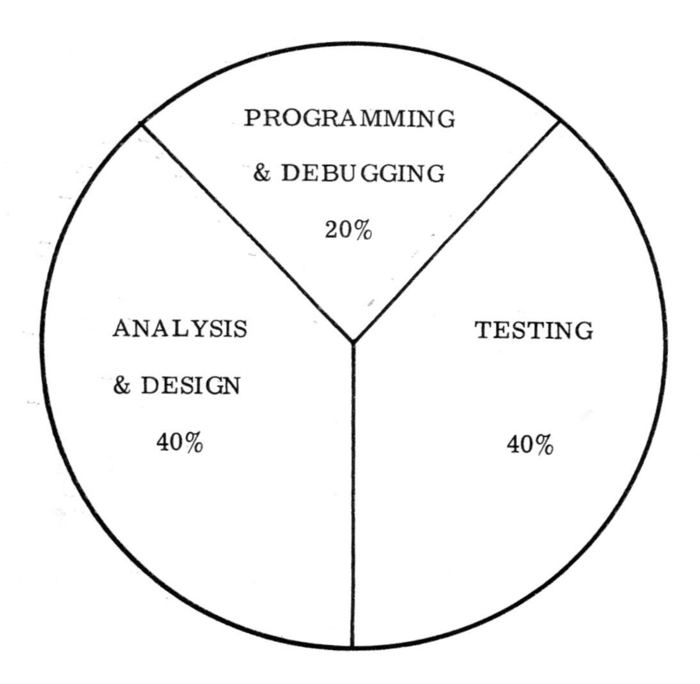

Fig. 3.9 Software development breakdown

Testing is normally carried out at several levels:

1. Single unit. Here the smallest units of software are tested individually and in isolation.

2. Multiple unit. Gradually the single units are grouped together to form larger units and tests here will give a first indication of successful implementation of basic functions.

3. Software integration. At this stage the interaction and interfaces between units is tested, and also confirmation of the actual function's performance is obtained.

4. Hardware/software integration. This is probably the most critical test because, for the first time, the successful interaction between hardware and software is under scrutiny.

5. Acceptance. Finally the user checks that the system aspects are correct and that the product works to his satisfaction.

Software testing is thus normally a 'bottom-up' procedure as described earlier. Other testing approaches are used, including 'top-down' and 'sandwich', which means starting from top and bottom and meeting in the middle, but bottom-up allows testing to progress in parallel with software development and is thus more akin to established equipment procurement procedures. Testing will also continue throughout the life cycle of the product, particularly when software enhancements are necessary as a result of changing or developing requirements.

Prototypes

In hardware engineering development of any kind it is normal practice to build a prototype in order to confirm the requirement. This approach can be used to good effect in software development. The use of a software prototype is an important method of reducing risk and enables resource estimates to be refined and provides a forum for any untried techniques to be evaluated. A prototype will also provide a visible, tangible model which can be demonstrated to the user in order both that he can confirm that the design does, in fact, reflect the requirement and that he can gain confidence and experience. It will also be important to start software development as early as possible in a project and a prototype will allow this work to be brought forward independently of other parts of the system.

The approach involves the building of a HOST computer to represent the eventual TARGET, or operational, computer in every respect. The host computer must be built quickly and will need a high level language and the necessary compilers and associated programs to be immediately available. The software environment, such as the interface software programs which control the peripherals, in which the prototype software is developed on the host machine should be indistinguishable from that in which it will eventually be operational. As soon as each portion of software has been through all its development and testing, it should be inserted into the simulated software environment on the host computer; in this way the software on the host machine will gradually evolve into operational software.

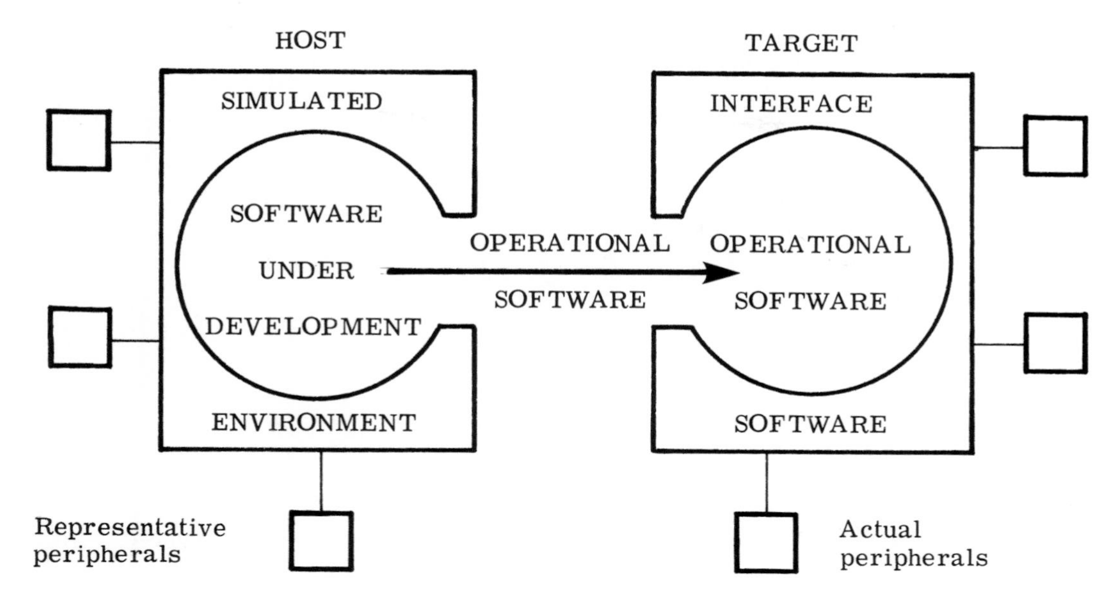

Fig. 3.10 Host/Target

Summary

As a result of the technical advances, better management, dedication of the programmers, well defined testing procedures and the use of prototypes, software products are without doubt improving in quality. The important point is for users to insist that quality assurance procedures are imposed and not to accept lower standards in order to save time and cost; in practice cost limitations often cause corners to be cut and quality assurance has in the past suffered as a result.

SOFTWARE PORTABILITY

PORTABILITY is a measure of the ease with which a piece of software can be implemented on a number of different machines. A program may justifiably be described as portable if the effort involved is small compared with that needed to rewrite it for each machine. Clearly such programs will be machine independent, or very nearly so.

In determining the best way of achieving portability in a particular application there are many factors to be considered, for instance:

How much will it cost to produce the first single implementation?

How much will it cost to produce subsequent implementations?

What is the range of these further implementations - how many different machines will be involved?

How efficient will these various implementations be - how important is this?

Perhaps the most obvious approach would be to make use of the universally available high level languages, FORTRAN or CORAL for example. The range of implementations would be very wide and, providing programming was limited to a standard subset of the various language dialects, the cost should be reasonable. The language chosen could however be extremely unsuitable for many of the algorithms or applications. Consequently the initial costs would be high as the program would be relatively laborious to write; it would almost certainly be inefficient.

What however if the range of machines does not have a suitable common language? The solution may lie in the use of a TRANSLATOR, a program which converts from one programming language to another, if such a translator is available. The complexity of such programs may, however, be so great as to render them far too expensive.

There is no clear optimum or universal solution to the problem; each case must be reviewed on its merits. The problem is a subset of the larger problem of attempting to produce a universal, omni-purpose high level language.

SOFTWARE COST ESTIMATING

The Problem

One of the most vexed questions which occurs in every software based project is that of how to estimate the cost of producing the software to carry out a particular task. There is no easy answer. The situation is aggravated in most cases by demands from the user for cost estimates before the requirements have been documented and fully explored or even while design is in its very early stages.

In simple terms the estimating of software follows the same rules that are applied in the development of any other item of equipment. The process of estimating the magnitude and timescale of a software task is one of progressive refinement right from the start of the task, through the requirements definition stage, and continuing until the software package is delivered to the customer and formally accepted.

The estimations involved are required to produce inputs to several project-wide activities including manpower planning, timescale estimating, cost control and budgeting, and integration of project-wide components. Early in a project very little detail will be known and estimates will be broad.

Macro-estimating

Macro-estimating is, as the name suggests, estimating on a global or large scale basis and is all that is possible before sufficient detail is available to enable detailed estimating to take place. Whilst this may seem to be rather more of an educated guess than a mathematical process there are a number of aids that the estimator can use which will become factors in his work:

1. The estimator's previous experience of systems of similar size and complexity.

2. The known degree of detail of the actual task.

3. The qualifications, experience, and level of understanding of the software team involved and thus how much training or how long a 'learning curve' will be necessary.

4. The time available.

5. The degree of complexity foreseen.

6. The number of novel design concepts and amount of 'new ground' to be broken.

Although this process is, by definition, a global appraisal of the situation, it should be possible to sift the available information and summarise it usefully to answer specific questions, such as "How much effort will be involved, in man-years"?, or "What is the estimated timescale, having considered all the factors above"?.

To provide the information to answer such qeustions and to provide a well documented basis for further iteration it would be necessary to record, too:

1. How were the first two answers arrived at, what are the critical parameters used, and how was each one evaluated?

2. What assumptions have had to be made?

3. What is the activity schedule used as the basis for the estimating and has it been agreed by all parties involved?

4. What do other people, like the software team, think of the estimates?

Having produced the first global estimates there must now follow, at regular intervals, progressive refinement of them. This will extend at least until the start of the software design stage. After that it should be possible for the software team to provide more detail of the actual task.

Micro-estimating

During software definition and design stages it should be possible to provide detailed estimates of such factors as:

1. CPU utilisation.
2. Store requirements and type.
3. Timescale for each software package development.
4. Activity schedules.
5. Requirements for personnel.
6. Requirements for hardware.

This type of detail would be used to refine the earlier macro-estimate. The progressive refinement should then continue. It would probably be based on regular reports by the individual members of the software team when progress and technical difficulties encountered can be taken into account.

SUMMARY

It is now probably clear that the development of hardware is the relatively easy part of devising computer systems. Software development is far more complex and far more of an intellectual challenge. Great strides have been made in meeting that challenge in the past decade but we are probably still at the beginning of what will become a very large and most important industry. It may, for example, become as important as publishing is today.

SELF TEST QUESTIONS

QUESTION 1 What do you understand by 'Low Level Languages'?

Answer ....

....

....

QUESTION 2 What do you understand by 'High Level Languages'?

Answer ....

....

....

QUESTION 3 What is the function of a compiler?

Answer ....

....

QUESTION 4 Do High Level Languages have any disadvantages, and if so what are they?

Answer ....

....

....

QUESTION 5 What would be the main advantage of the advent of a single ideal all-purpose High Level Language?

Answer ....

....

QUESTION 6 Why is it important to specify test data early in software development?

Answer ....

....

....

QUESTION 7 What is the purpose of multi-programming?

Answer ....

....

....

QUESTION 8 What is the function of an assembler?

Answer ....

....

QUESTION 9 Why is there a need for software tools? What sorts of aids and tools are available?

Answer ....

....

....

....

....

QUESTION 10 What are modular design and structured programming?

Answer ....

....

....

....

ANSWERS ON PAGE 198

4.

Development of a Computer System

INTRODUCTION

Information

Having looked in general terms at how computers are used and in more detail at software and hardware in Chapters 2 and 3, it is worth considering the way in which a computer system is developed and is brought into service. Digital computer systems can be described as information systems and this chapter deals particularly with the implementation of such information systems.

Good management has been defined as 'the ability to make the right decision on inadequate and late information from dubious sources'. Indeed after everything has been said that can be said about scientific management, it remains true that nothing can relieve the manager of the job of making the final decision, and the 'hunch' must play its part, for this is the art of outstanding management.

> "Our stability is but balance, and wisdom lies in masterful administration of the unforeseen".
>
> R. Bridges (The Testament of Beauty)

Yet much can be done to reduce the uncertainty in decision making, and Methods Engineering and Operational Analysis are two of the many means of reducing this uncertainty. Such methods must provide management with information that is accurate, that is to say factually correct; it must be concise because too much detail may not be cost effective or manageable; it must be understandable, which means that it must be represented clearly; it must be timely, current or up-to-date and economic because the value of the information must justify its cost. From these requirements it can be seen that the manager's interest in the computer is in its potential to provide better information.

SYSTEMS ANALYSIS

In Chapter 1 the importance of considering the computer as a system, or to be more precise an information processing system, was stressed. The general definition of a system said that any system is a sub-system of a larger system. This is clearly true for the computer; it is a system in itself, yet it is inevitably part of the larger system it supports. The activity of examining whether a computer is required, and if so, whether information can be processed at an acceptable cost, then how the computer system may best be matched to the system it supports, is described as SYSTEMS ANALYSIS.

It must be emphasised that there is nothing new about systems analysis, it is a technique that man has used, though not recognised by such a grand title, to improve his conditions ever since he formed into social groups. It is a technique from which many of the modern management methods originate, a technique which some prefer to describe as 'commonsense'.

Implementation Problems

The introduction of a computer into an organisation, and the transfer of data and procedures to that computer may pose special problems. First the high, sometimes catastrophic, cost of mistakes. Then the fears of departments that their previous autonomy will be destroyed as their jealously guarded files and procedures are investigated, extracted, changed and computerised. And, of course, the enforced and often resented inter-departmental consultation and exchange of information.

For these reasons the correct analysis of the present situation is vital. Then consideration of changing the system to suit computerisation is essential. This is the role of the Systems Analyst.

Sequence of Analysis

As with all management situations systems analysis follows the steps on the well worn path of:

specifying the problem

recording the facts

examining the facts

developing the best solution

installing the solution

monitoring the result.

Thus systems analysis can involve a wide spectrum of activity. Some of these areas indeed appear to fall outside the accepted definition of analysis: the last three are strictly design work. However it is essential that these phases are linked so that the analyst has real involvement in the design phase, if only to avoid the suggestion of idealistic but totally impracticable proposals.

The Problem and the User

At this stage the involvement of the user himself in systems analysis must not be overlooked. It is the user who defines the requirement for analysis, though in the first instance this requirement may be unavoidably vague. Hence the purpose of the first activity of specifying the problem is to define the requirement clearly and accurately. This may seem rather an obvious first step in any undertaking, but experience shows that there is more than one large computer project in which the requirement has not been sufficiently clear, and the end result has been an extremely expensive white elephant.

To specify the requirement, clearly and accurately, may necessitate an extensive preliminary study, one which involves some limited consideration of the other phases of systems analysis. This is acceptable, but it is essential that the user agrees the requirement that emerges from the first phase of analysis, before any other phase is considered in detail. It is for the user to decide whether the requirement is right, and whether further analysis is warranted for it may well be that, as a result of the first phase, it becomes clear that ADP is not required, or is too expensive and that the solution to the problem lies elsewhere.

Developing the Solution

It is difficult to separate the next three phases, which are interlocking. Developing the best solution will necessitate frequent recording and examination of the facts. Clearly the user must ensure that the analyst has proper access to the facts, and the important question is who decides which is the best solution? This decision must rest with the user and not the systems analyst. How good any solution put forward by the systems analyst is, will depend not only on his competence but on how carefully the user has been able to define the requirement. No matter how well the requirement is defined, the analyst will tend to produce an ideal system; it will be one which best suits the machine in relation to the overall requirement.

Dangers

There are two dangers in accepting such an ideal system: the first stems from the fact that the computer never achieves complete automation, men must still exist somewhere within the system, and they are often the same men who previously performed the task by entirely manual means. If a new system is to work, then the confidence of these men must first be gained. There is often, in effect, a psychological barrier, a resistance to change, which might not be recognised as a relevant factor by the systems analyst but which should be recognised by the practical user. In such cases it may be far wiser to introduce a simple system first and so gain the confidence and support of the men who will run the system rather than jump straight to the 'ideal' solution.

The second danger is that the ideal solution will probably incorporate too many untried techniques. Without new techniques there is no progress, but a fine balance must be struck between those new techniques which are vital, or offer significant benefits, and those which may be aesthetically pleasing to the analyst, but realistically offer little more than a potential source of delay and additional cost.

Safeguards

It is for the user to ensure that these hazards are avoided and this can be done in three ways: first by ensuring the involvement of the user throughout and thus that the requirement is correct and incorporates all known constraints. Second, by tasking the systems analyst to produce not just one 'best' solution, but a number of options over a carefully selected range; and third, by ensuring that whatever solution is chosen the systems analyst is, and knows he will be, involved in the last two activities which are installing the solution and monitoring the result, or rather making the solution work.

THE DEVELOPMENT CYCLE

The Phases

Most large ADP projects are envisaged as consisting of five phases. A preliminary study, then a full system study which is the basis for approval; once approved, the system is implemented, and finally enhancements and modification are part of a post-implementation phase.

Preliminary Study

The preliminary study is a phase devoted to assessing the desirability and feasibility of applying ADP to the area of activity covered by the project and to give a specific object to later phases. This phase is often called the FEASIBILITY STUDY: it may be an inappropriate term because the feasibility of applying ADP is in most cases not the main question. The important question is "is it desirable?" In military systems the Feasibility Study is normally carried out as the result of a user sponsored requirement.

At the end of the preliminary study a report is normally produced which can be scrutinised by users, sponsors, technical authorities and project management. If the report recommends that the project should proceed then a larger team is recruited for the full system study phase.

Full System Study

A full system study embraces the majority of the work of systems analysis and will be described in detail. It can be sub-divided into a number of activities:

1. Establishing starting points.

2. Investigation.

3. Analysis.

4. Outline Systems design.

5. Study interactions with other subsystems.

6. Detailed system design.

7. Report.

The order given represents the general progress through the full system study phase but the activities do not necessarily occur in rigid chronological order. Looping back is frequently necessary. For example, the process of analysis often reveals the need for further investigation.

Establishing starting points involves establishing or clarifying the aim of the study, which organisational frameworks are involved, any applicable constraints and the production of a task plan.

Many books on systems analysis are weak on the investigation aspect. In particular they often imply that this phase involves simply the investigation of an existing system, implying that a manual system carrying out all the functions of the system being designed already exists. This is not, in general, true. Investigation involves the examination of the following to an extent varying with the particular application:

> Users' views and requirements for the new system.
>
> The 'Object System', which is the system the information system is to model. It might, for a stock control system, be the actual people, warehouse and goods; or, for a Command and Control System, the staff officers, staff cells and the military units under command.
>
> The existing information system where this exists. This will comprise, for a manual stock control system, such items as the issue notes, receipt notes, transaction records and ledgers; for a Command and Control System the type of items will be the message pads, proformae and maps.

The main sources of information in this activity are people, documents and direct observation. Efficient documentation during this phase is essential as it will form the basis of all future work; if an item of information or activity fails to be documented it will be excluded from further consideration. Aids to documentation include organisation charts, establishment tables, specification of clerical documents and files, relevant technical and operational manuals, flow charts, multiple activity charts and reports of discussion.

The task of documenting the current manual system can be a lengthy, costly procedure. Many individuals do not have a clear idea of what they do, or how they fit into a system, and even if they do, they may have difficulty expressing it.

Analysis is concerned mainly with the existing system and the tasks of first, completing a model of the existing system by piecing together the data provided by the investigation and, second, the critical examination of the model.

The analysis activity should produce answers to a number of important questions. For example, "How well is the present system doing?" In particular "Are the outputs needed?; are there bottlenecks in the information flow?; can duplication of effort be eradicated?" Then, "Do the terms of reference call for a 'cure' to a 'disease' different from that which actually exists?", and, "Are all the functions of the system necessary?" Occasional analysis of a system even without the aim of future automation is often, of course, very cost effective and a useful management tool.

Outline systems design is possibly the most important activity and it is all too often skimped by those wishing to charge ahead with the detail. It is almost impossible to separate this process from the study of interactions with other systems and in military areas the interoperability constraints are numerous. There is a need for a repetitive process of synthesis and analysis. A model is created, critically examined, improved and this process is repeated until an apparently adequate design has been achieved. This is normally unlikely to result in a simple automation of current manual processes. From this activity it should be possible to decide what the new system is to do, and an outline of how it is to do it; it is also most important to document the arguments behind these decisions.

It is important that all possible alternative approaches are exposed. They must be considered in sufficient detail to allow the weaker ones to be discarded after offering the outline design for comment and criticism, but before getting involved with the detail. A large number of factors will have to be considered and the importance of this phase cannot be overemphasised.

A study of the implications of interactions with other systems may be critical in a particular application; for example a microprocessor in a tank sight may have to interface with an infra-red device, a laser and a loading unit; these are all subjects with their own individual design. A command and control system may have to interface with trunk radio, VHF combat net radio, and other ADP systems which may all be at different stages of development or built to different technical standards. Some more detail of the interoperability problems in military equipment can be found in Chapter 8.

Some general considerations in detailed system design will probably be:

the importance of flexibility and balance,

the avoidance of reinventing the wheel,

the importance of modelling and testing exactly to the specification, and the use of the exception principle where only situations deviating from expected standards are reported.

One very important decision to be made at this stage is the distribution of processing power and database. There are several choices from a monolithic system shown in Fig. 4.1 to a fully distributed system in Fig. 4.2.

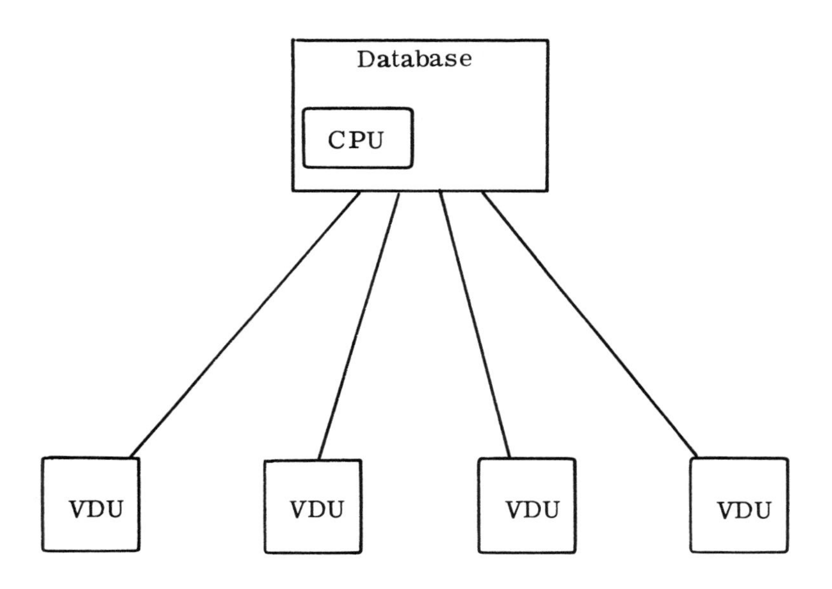

Fig. 4.1 A monolithic system

A monolithic system has a single centralised processor and database with terminals such as VDUs and teletypes attached albeit over extended communications links. Such a system could be vulnerable in that a failure, for whatever reason, of the processor would render the complete system inoperative. In a peacetime environment this could be bad enough but in war it would be unacceptable. A fully distributed system has processing power and a complete copy of the system database at each node of the system; nodes are connected by communications links and an amendment to the database of one node will be automatically transmitted to all other nodes. The failure or loss of a node will not seriously degrade the functioning of the remainder of the information system.

In practice it is possible to strike a balance between these extremes; for example, in a command and control system it may not be necessary for all HQs to hold a complete copy of the database. Care taken to identify the critical data holdings can reduce the costs of the expensive store. Intelligent terminals, that is terminals with some processing power, can handle messages and provide displays without recourse to a centralised processor; such peripherals can be used to good effect in balancing the design.

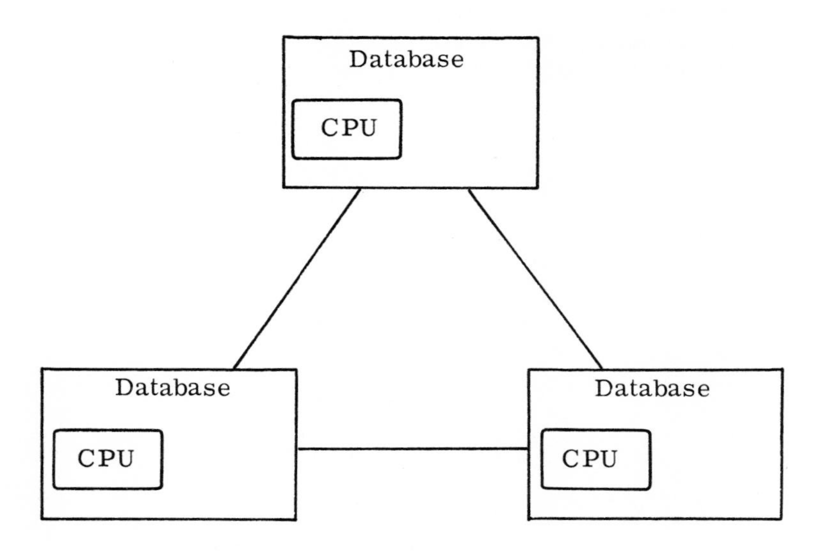

Fig. 4.2 A distributed system

The general procedure is to design iteratively at increasing levels of detail starting from the top and working down. This is what has already been described as 'top-down design'. A general division and flow of work is to consider first the outputs required, then the inputs available, followed by the procedures, databases and files which will be necessary. A detailed System Design will probably include the definition of the outputs required, the input data required to yield these outputs, the information to be held in files, and the computer procedures linking inputs, outputs and files. It should also specify security and staff requirements, the foreseen organisational changes and the manual procedures required to support the new system. A top-level 'software plan' will be produced and then a definition of hardware requirements; this does not imply the selection of a particular manufacturer's equipment, but rather is a statement of the hardware capabilities required. As a result of all the work involved in this activity an implementation plan can be drawn up. There will be numerous documents produced and management of these documents alone will be vital to the success of the project; strict configuration control is essential, as mentioned in Chapter 3.

At the end of the full system study a report should be prepared and must include the system design and costs. It should also include the cost of an improved manual system, where appropriate, for comparison purposes and set out recommendations.

Approval

Once the full system study report has been produced it is normally scrutinised by various authorities. Normally they are the user or sponsor, project management and, in government or military applications, the technical authority. The systems analyst will be active during this phase in giving presentations and briefings and generally selling his system.

There are three important decisions which must be taken during this phase. First, whether the recommendations in the report should be accepted or other selected alternatives should be chosen. Then, if the project is to be implemented, how the facilities required are to be provided. In particular in military systems it must be determined whether the requirements could be met by using hardware/ software provided from some other in service equipment. Finally, any necessary organisational and operational changes must be planned.

Implementation

A number of activities are progressed in parallel during this phase. If additional or new equipment is required, it must be developed or procured; in military equipment this will be to meet an official 'Operational Requirement' based on the facts prepared in the full study report. Communications requirements must be fulfilled, user staff must be indoctrinated and, where necessary, trained. For an operational military equipment to go into the field there will be the added problems of redesigning vehicle installations to accommodate the hardware. It will also be necessary to study the increased security requirements, which will involve both the physical guarding of equipment containing classified information and the protection of such information from unauthorised users. A major task will be that of tests and trials, which could last several years.

Post Implementation

Once implementation has been completed there should, in theory, be a sharp run down in staff concerned with systems analysis and programming. For various reasons, not all of them entirely creditable, this often does not occur. One important remaining task is to modify the system in the light of practical experience just as in the case of all new equipment. After a system is implemented, more checks should be made to see that it actually functions according to the requirement: it can lead to embarrassment if unrealistic claims are made at the full system study stage. It is, however, accepted that during development not all the possible combinations of programs can be tested and that faults may occur later in untested areas. The technique of detecting, diagnosing and correcting errors, or BUGS, which may occur in hardware or software, is known as DEBUGGING and will continue long into the service of the computer.

The role of the Systems Analyst changes during the life cycle of a project. During the feasibility study and the early part of the full system study he is an investigator; he then becomes a designer. During the approval phase he becomes, to some extent, a salesman. During implementation he is a coordinator and progress chaser.

FLOWCHARTING

Systems analysis includes the detailed examination of a process or system, breaking it down into a series of steps, and finally documenting a description of those steps. It is essential, as has already been mentioned, that at all steps of this process there should be good communications between all those involved, and in particular between the analyst and the user. Such communication is normally based upon an established technique called FLOWCHARTING.

A flowchart is a pictoral method of representing a process in a logical progression of steps. It does not necessarily describe a computing problem but it is a method commonly adopted in system analysis and in computer programming.

Flow charts may be drawn to varying depths of detail. At the higher level they are designed to delineate problems and are known as SYSTEMS FLOWCHARTS. They describe steps in general terms such as 'Sort cards into suits' or 'Sort vehicles into types'. At the lowest level a PROGRAM FLOWCHART is orientated to a particular programming language and will break a process down into the smallest possible steps; each step is numbered and equates generally to a single instruction. For example 'is the card a spade?' or 'is the vehicle a tank?'.

Drawing the Flowchart

There is no universal standard of flowchart conventions but those shown in Fig. 4.3 are commonly used and understood.

There is only one start and one stop in a process, though getting from one to the other is not necessarily achieved by the most straightforward path. Individual steps which are actions are described in rectangles. Individual steps which involve taking decisions are described by writing the associated question in a diamond shape. Note that the questions must be reduced to their simplest form, that is a question allowing only one of two answers such as, 'yes' or 'no', or 'true' or 'false'; but there must be no allowance for a third option such as 'maybe'.

Most computer programs have constructions which either involve repeating certain parts of the program or leaving certain steps out if a particular condition applies. Flowcharts follow a similar pattern with alternative paths stemming from a decision box. One path follows the normal sequence of events whilst the alternative will either jump some actions or loop back to a previous action. Although there is only one end it may be reached by various different paths and it is therefore acceptable to have several alternative STOP symbols.

Figure 4.4 shows an example of a flowchart for a non-computer problem and illustrates the use of all the conventional symbols (except, of course, the data input or output symbol).

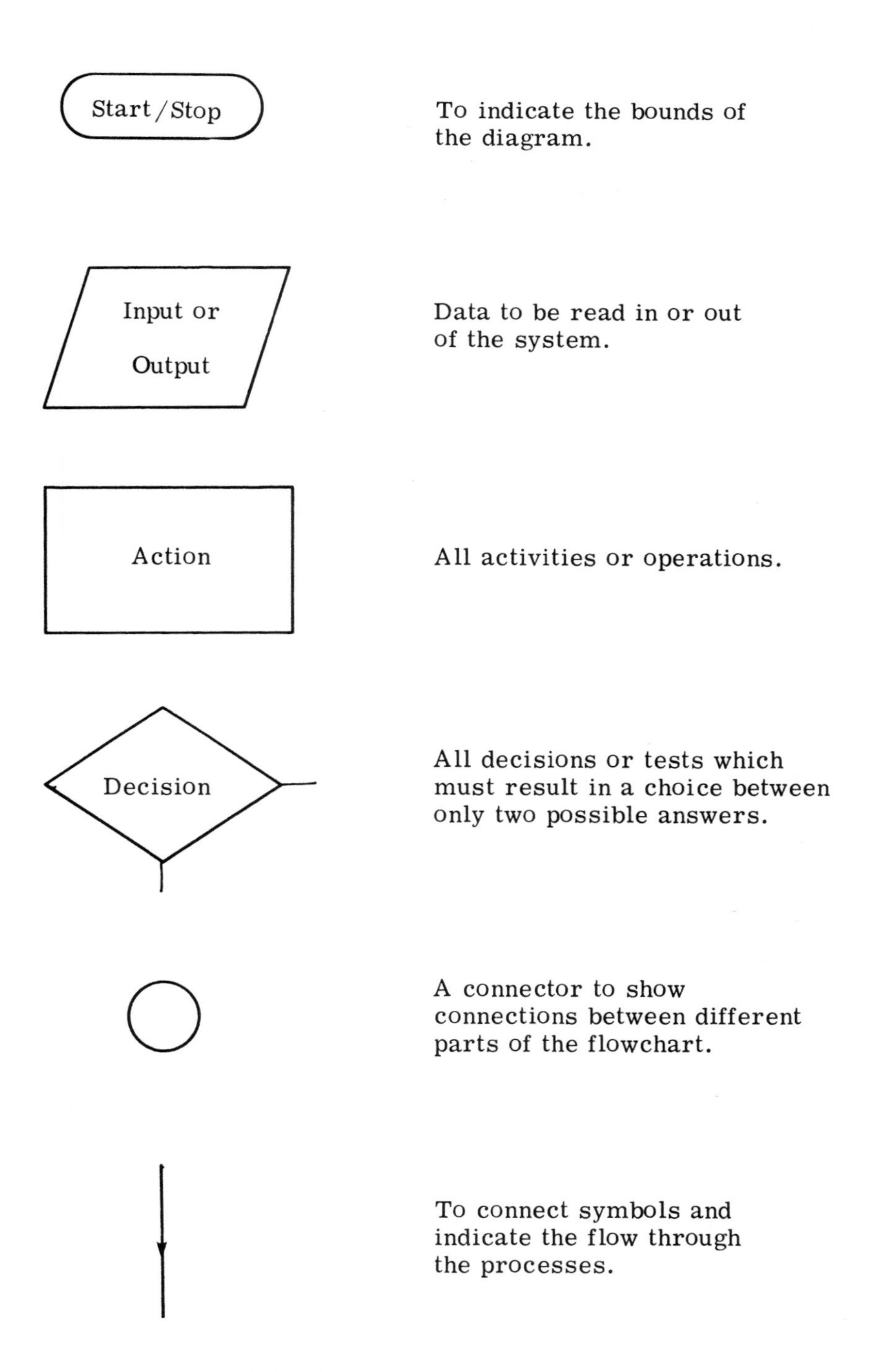

Fig. 4.3 Common flowchart symbols

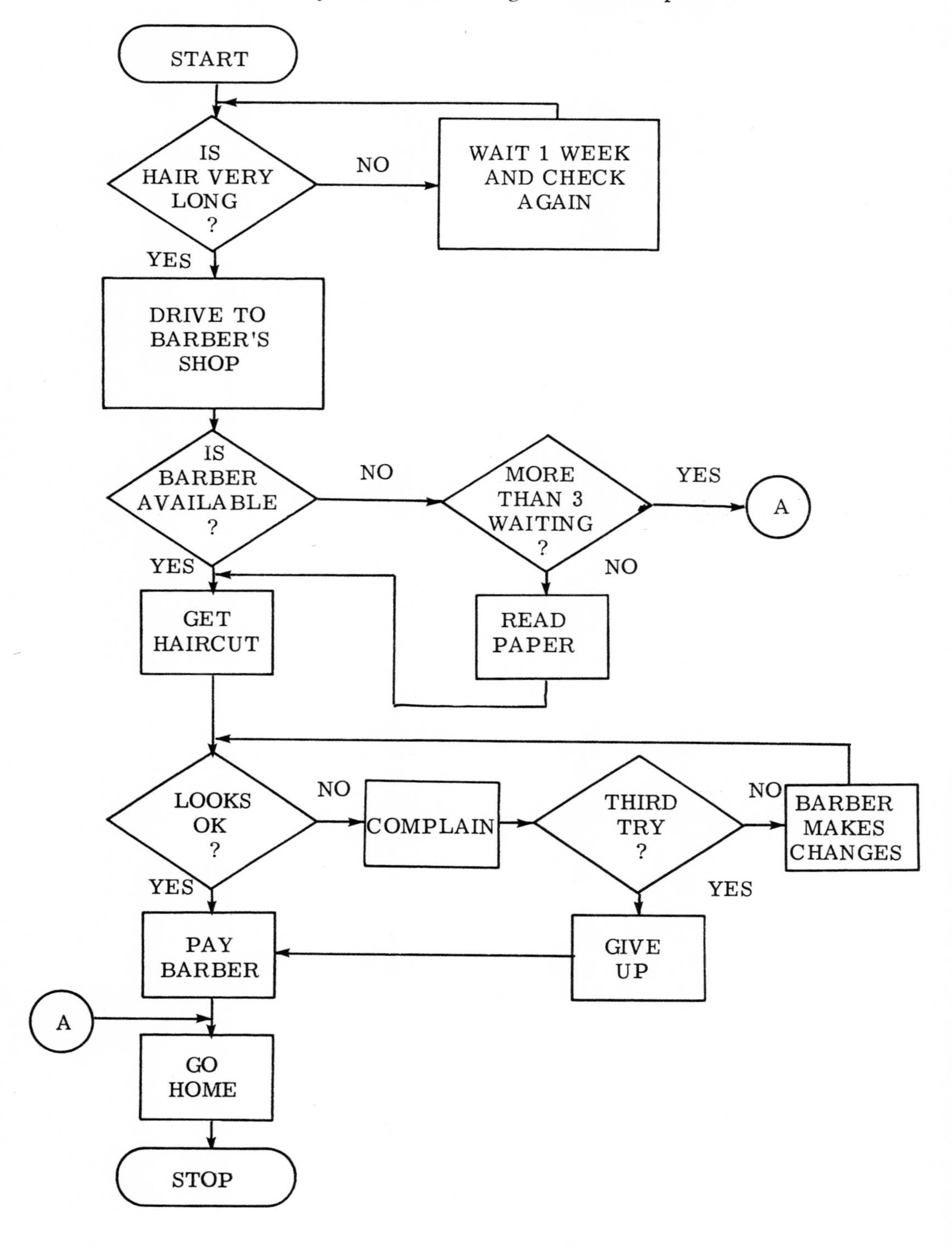

Fig. 4.4 Non-computer problem flowchart

An example of a computer systems flowchart is shown in Fig. 4.5; the problem is to input three numbers representing the length, height and width of a box into a computer in order to output the volume and surface area if the volume exceeds 10 cubic metres.

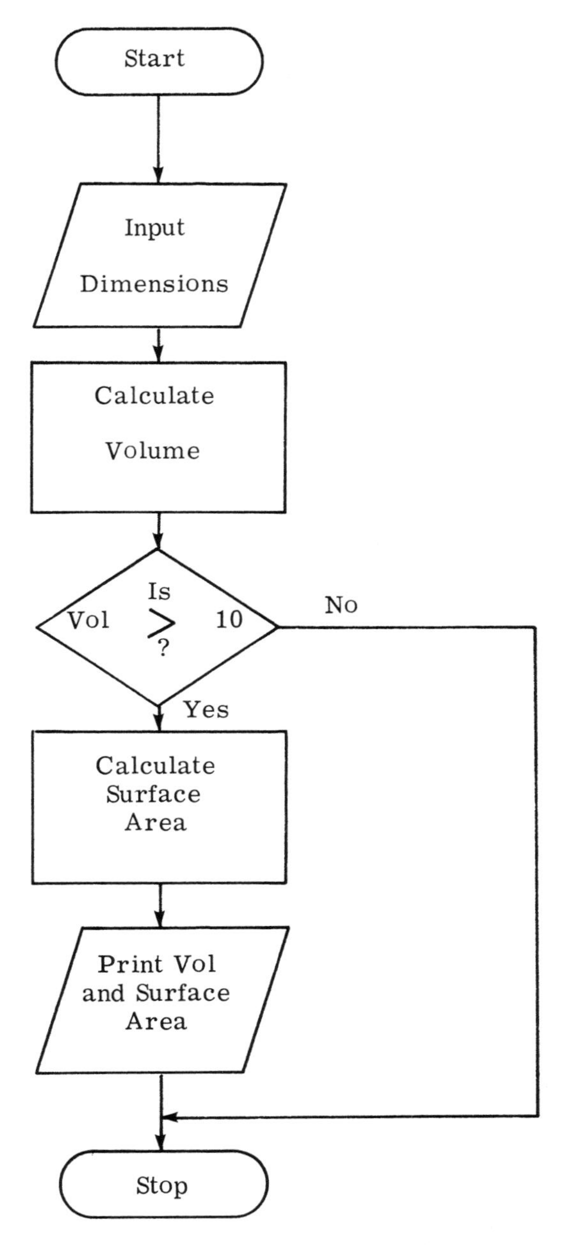

Fig. 4.5 System flowchart for the box problem

A program flowchart for the same problem at a more detailed level is shown in Fig. 4.6. Note that the steps are very small, clear, and logical.

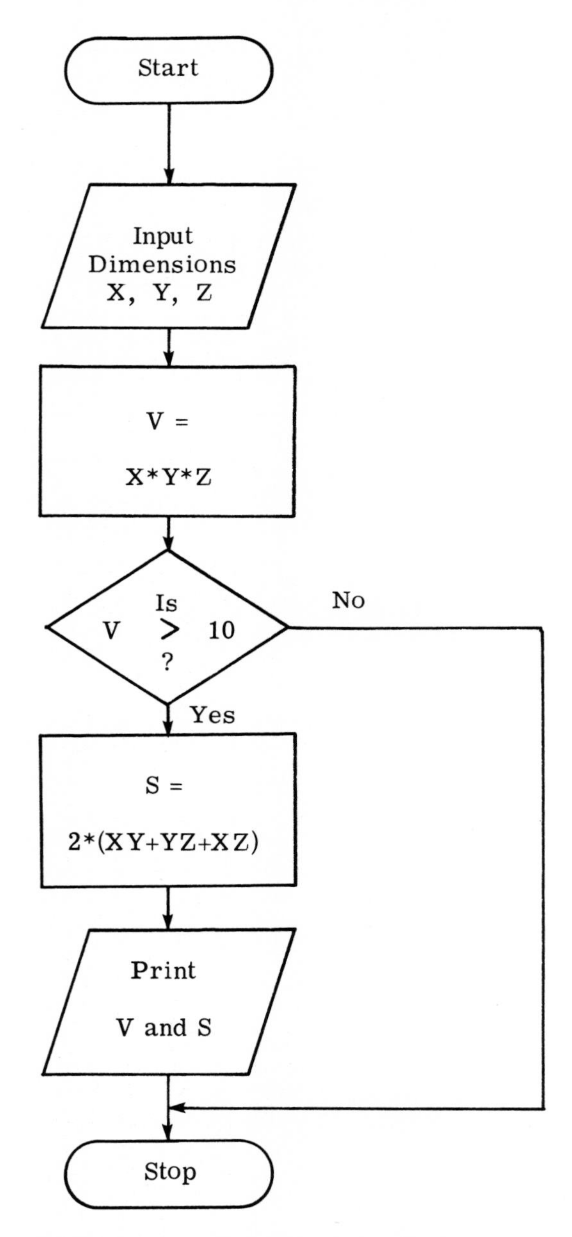

Fig. 4.6 Program flowchart for the box problem

Two further examples are shown in Figs. 4.7 and 4.8.

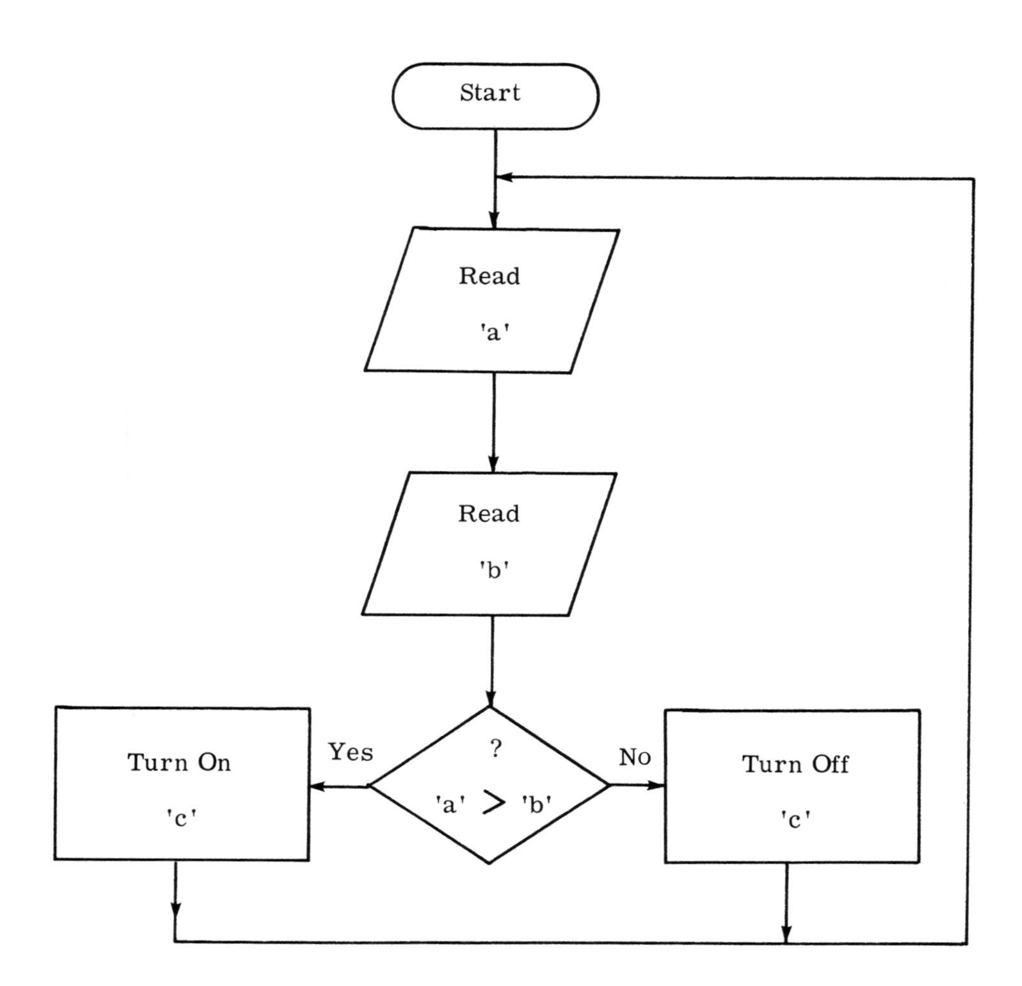

Fig. 4.7 The systems flowchart to illustrate the procedure to monitor continuously two meters a and b, and then turn on a switch c if a > b, otherwise turning switch c off. Note that in this example, since the procedure is continuous once it has started, there is no STOP box.

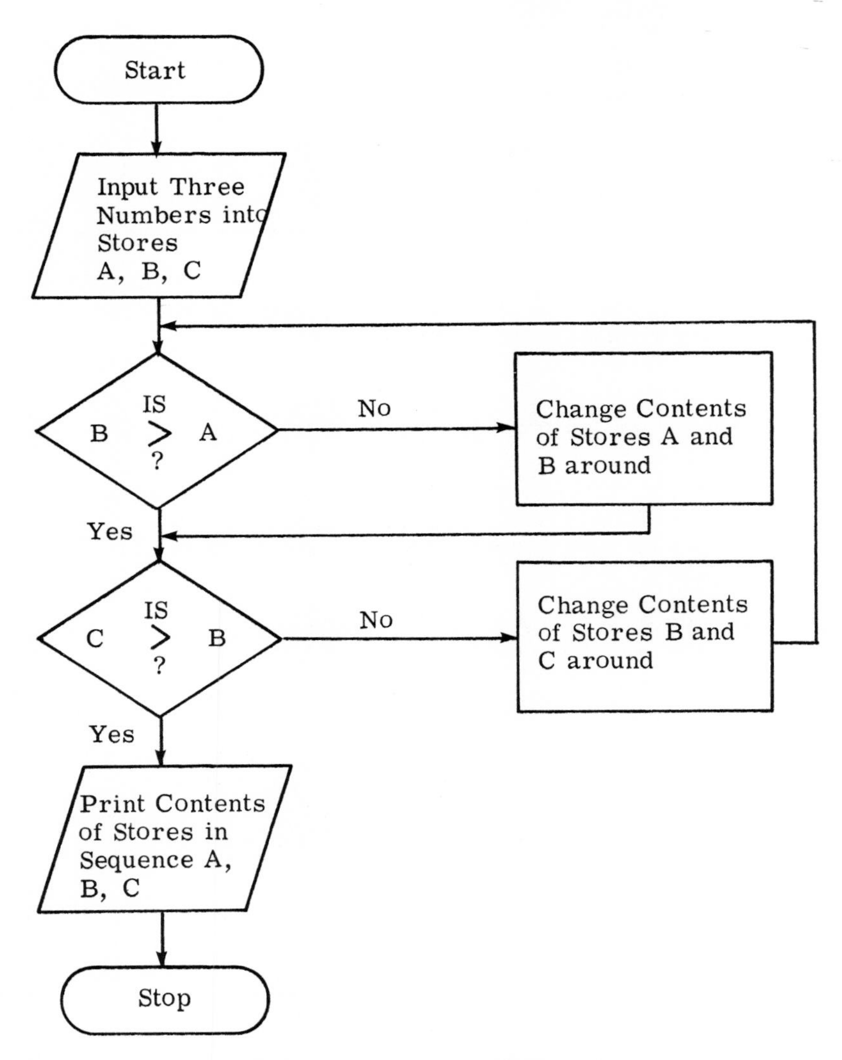

Fig. 4.8 The systems flowchart to illustrate the procedure to look at 3 numbers and print them in ascending order. Note that the procedure used by the computer to change the position of the pairs of numbers may be quite complicated and the program flowchart for this example could have many more boxes.

The transistors, resistors and capacitors forming the circuit are produced by first using photographic masks to define the different areas required for each component, and then passing hot gases over the surface of the wafer in order to produce regions of controlled impurity in the silicon, or to grow different layers on the surface. Typically, each transistor will occupy an arêa only a few micrometres square.

The equipment required to carry out this processing is very expensive, and the production yield is usually very low. Often only one or two percent of circuits pass their tests, so quantity of production is the key to the low final cost. A silicon wafer containing over two hundred and fifty complete circuits is shown in Fig. 5.1. After testing, the wafer is cut up, the faulty circuits are discarded and the good ones individually mounted in their packages. Connections to the package leads are made by me ıns of gold bond wires which are welded to pads at the edge of the circuit. Figure 5.2 is a close-up of a microprocessor circuit showing the bond wires connecting it to its package, and Fig. 5.3 shows the same circuit mounted in its complete package but before the final cover seal has been attached.

Fig. 5.2 A microprocessor circuit

The packages provide a hermetic seal, preventing contaminants from reaching the circuit and affecting its operation. Packages are either moulded plastic, or they can be ceramic which are suitable for use over a wider temperature range. The packs are often referred to as dual-in-line (DIL) because there are two rows of connections, one either side of the pack. Whilst the final size of the fragment of the silicon wafer is usually no more than 7 mm square the packs are much larger in order that the final connections appear at a spacing that can easily be catered for by a printed circuit board. The pack in Fig. 5. 3 has forty connecting pins, twenty per side at 2. 54 mm (0. 1 inch) spacing, and the two rows are 15. 24 mm (0. 6 inch) apart, so the size of the final package is about 51 mm by 15. 5 mm.

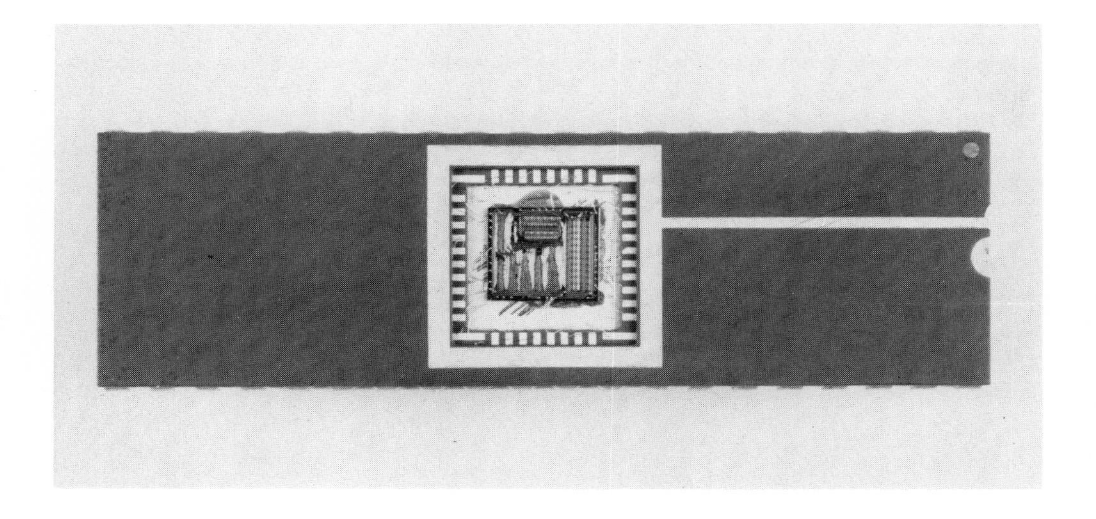

Fig. 5. 3 A packaged microprocessor

Currently, integrated circuits are available containing up to 150, 000 transistors, whilst prototypes with 250, 000 or more have been produced. Despite the internal complexity of these circuits they are available at low cost because of their mass production; their reliability is due to the almost complete elimination of soldered joints and connectors.

THE MICROPROCESSOR

Main Components

The microprocessor, or MPU, consists of the major elements of a computer CPU in a single integrated circuit. The structure of a typical MPU is shown in Fig. 5.4 The Program Counter (PC) is the key to the operation of the MPU. The MPU is connected to a store, or memory, containing the program to be executed. This connection is first by the address bus which carries the numerical address of the memory location which the MPU wishes to access; then there is the bidirectional data bus which can carry data from the memory to the MPU or vice versa; and finally there are a number of control lines which ensure synchronisation between MPU and memory.

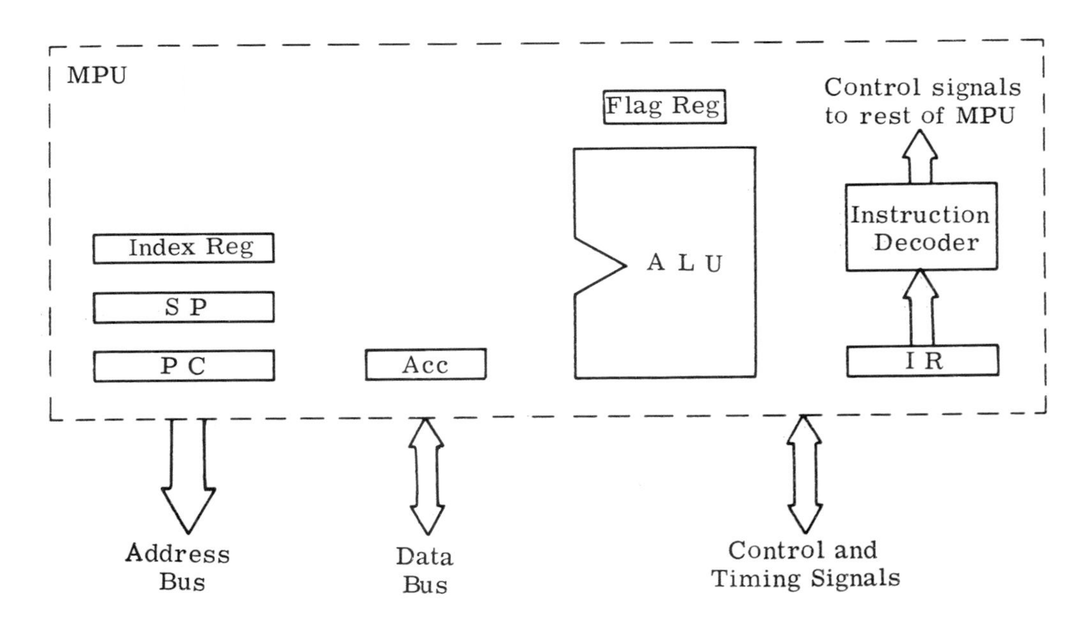

Fig. 5.4 A typical MPU

Operation

Initially the PC contains the address of the first instruction in the program. This address is placed on the address bus and the memory responds by copying the data stored at that location (the first instruction) on to the data bus. The MPU reads the instruction in binary code and places it in its Instruction Register (IR). From there it passes to the INSTRUCTION DECODER, which then produces the signals which cause the MPU to execute the instruction. The PC is then incremented by one, so that the next instruction may be read and executed, and so on. The MPU is thus alternating cycles when it fetches instructions from memory, with cycles when it executes them. MPU operation may therefore be summarised as:

Fetch instruction pointed at by PC

Execute instruction and increment PC

Fetch next instruction etc.

The other registers shown in the MPU are concerned with the execution of the program instructions. The ACCUMULATOR is the main data storage register within the MPU. It is used to hold a binary number which is about to be operated on by the ALU, or which is the result of a previous operation. Thus if a section of program consisted of the binary codes for:

Load accumulator from location 29

Add location 30

Store accumulator at location 31

then the operation of the MPU would be as follows. The PC holds the address of the instruction "load accumulator from location 29". The instruction is fetched and decoded and causes the MPU to place the binary number corresponding to 29 on the address bus. The number stored in location 29 is then copied, via the data bus, into the accumulator. The PC is incremented and the next instruction fetched. The MPU then puts the address 30 on the address bus and copies the content of that location into the ALU, where the number held in the accumulator is added to it. The result is placed in the accumulator, overwriting the previous value. The PC is incremented again and the next instruction fetched. The MPU puts the address 31 onto the address bus and copies the contents of the accumulator onto the data bus. The memory then takes this value from the data bus and stores it at location 31, overwriting its previous contents.

The types of operations described so far enable the MPU to work through a program step by step in a completely fixed manner. However sometimes it is useful to be able to arrange for the MPU to carry out one of two courses of action, the choice will depend upon some previous result. Thus we might wish the MPU to perform one set of actions if the result of a subtraction were zero, but another set if it were non-zero. The means of implementing this form of conditional instruction makes use of the FLAG REGISTER. This register is made up of a number of flags (bistable elements) which are set 'true' (1) or 'false' (0) depending upon the result of the last operation carried out in the ALU. For example, one of the flags will be a 'zero flag', which will be set true if the last result was zero, but otherwise set false. Corresponding to this flag the MPU will have two instructions, one being "jump if zero flag true", the other "jump if zero flag false", each followed by an address. If the condition specified is met then the MPU loads its PC with the address given and thus jumps to that point in the program. If the condition is not met then the MPU carries on executing instructions in strict sequence. In this way the processor can carry out different sections of program depending upon some earlier result. There are a number of flags to detect other events, such as positive or negative results, and these may similarly be used to control jump instructions.

The other two registers shown are the INDEX REGISTER (IR) and the STACK POINTER (SP). The stack pointer holds an address which points at a location in memory. If the contents of the accumulator are 'PUSHED' onto the stack they are stored at the address which is being indicated, and the SP is then arranged to point at the next memory location. If the accumulator is pushed again it will be copied into this new location and the SP will then point at the next location to that one. On pulling or 'POPPING' data from the stack the last item stored on the stack is copied back to the accumulator and the SP adjusted to point at the location of the one from last data stored, and so on. Thus the SP enables a Last-In-First-Out (LIFO) store, or stack, to be formed. This feature can be very useful when varying, and often unknown, amounts of data must be stored.

The index register also contains an address, and can be used to generate different addresses under program control. In the instructions described previously in this chapter the actual addresses (29, 30, 31) were directly specified in each instruction. An index register might instead have been used to provide so-called indirect addressing. The example used previously would then have become:

Load index register with the value 29

Load accumulator from address given by index register

Increment index register

Add location given by index register

Increment index register

Store accumulator in location given by index register.

This facility of indirect addressing is a powerful programming feature of great use in many applications.

In practice most real microprocessors contain a number of other registers, either intended for use as general purpose stores, or extra index registers or stack pointers. The discussion so far has avoided details of the size of the registers and buses (eg 8 bits, 16 bits, etc), and is quite general. However, MPUs are usually classed by the number of bits of binary data that they can operate on; this is termed their WORD LENGTH. An 8 bit microprocessor will have an 8 bit data bus, and will work directly with 8 bit numbers. If larger numbers and greater accuracy are required then 16, 24, or more bit working can be carried out by breaking the numbers up into 8 bit sections (bytes).

Processing Capability

Most of the earlier, and therefore most widely used, MPUs are 8 bit devices, and can only carry out simple arithmetic tasks such as addition and subtraction.

Other operations such as multiplication are performed by sequences of addition etc. The newer 16 bit MPUs can perform multiplication and division directly, and even more powerful 32 bit machines are beginning to appear.

If 8 bit microprocessors were to use 8 bit address buses as well as data buses then they would be limited to 2^8 or 256 memory locations; this would be insufficient for most applications. Consequently they normally use 16 bit address buses providing 2^{16} addresses and thus 65,536 different locations. Note that in computing terminology 2^{10}, which equals 1,024 is referred to as 1k; therefore 2^{16} is referred to as 64k since 2^6 equals 64. 8 bit processors access one byte of data at each memory location, thus a 16 bit address provides a storage capability of up to 64k bytes. The 16 and 32 bit microprocessors can address up to 16M bytes or more of memory, where 1M equals 2^{20} or 1,048,576.

Speed

The speed with which operations are carried out is governed by the frequency of a crystal-controlled oscillator, or clock, which provides timing signals for the MPU. Early microprocessors were limited to maximum clock rates of 1 or 2 MHz, but newer MPUs can work with clocks running at 10 MHz or more. The time taken to perform an operation such as the addition of two 8 bit numbers may therefore vary between about 0.5 μs and 5 μs, depending upon the MPU used. These are long times compared with those achieved by modern versions of traditional computers. It is because the main design aims for MPUs have been small size and low power consumption, and these have not been compatible with achieving high speed at the same time. Nevertheless the speed of operation of microprocessors is still more than sufficient for a very wide range of applications.

When greater speed is required, but it is still wished to take advantage of the small size of LSI and VLSI circuits then BIT-SLICE processors, sometimes referred to as bit-slice microprocessors, may be used. Rather than the whole CPU being contained in one integrated circuit the bit-slice devices provide a 'slice' through the CPU in each circuit. A 4 bit bit-slice circuit comprises 4 bits of the ALU, 4 bits of the accumulator, and so on. If a 16 bit processor was required then four of the 4 bit slices would be used. The improvement in speed occurs because the design aim of the bit-slice circuits was speed of operation rather than packing density. Work is also being undertaken on the design and manufacture of specialised, rather than general purpose, high speed LSI and VLSI circuits for use by the U.S. Department of Defense, and is known as the Very High Speed Integrated Circuit (VHSIC) programme.

THE MICROCOMPUTER

The Need

Only the microprocessor itself has been discussed so far, and whilst it is a vital part of a complete system it is not the whole of that system. In order to produce a complete microcomputer, based on a microprocessor, memory must be provided to store the program and data, and input/output devices, or INTERFACES, are required to enable the system to communicate with the outside world. The I/O connections are often referred to as PORTS. The structure of data and address buses provided by the MPU enables all these elements to be attached as desired. The block schematic of a complete microcomputer system is shown in Fig. 5.5, and the way in which the buses interconnect the components is clearly visible. In order to add extra memory or I/O devices, up to the addressing capability of the MPU, it is only necessary to connect them to the existing buses.

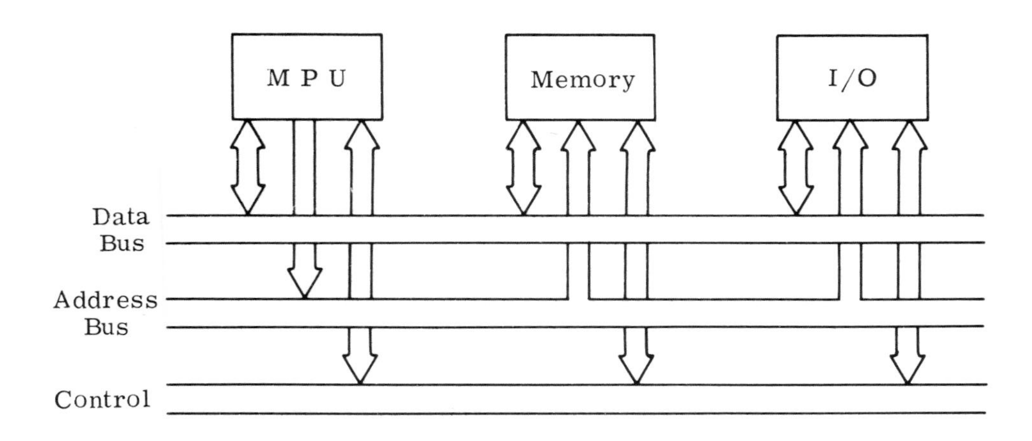

Fig. 5.5 A microcomputer

In contrast to most conventional computers, microcomputers are often used as dedicated systems performing a fixed task and actually embedded in some piece of equipment. It is therefore usually necessary to hold the program in a part of the memory which cannot be accidentally overwritten by the MPU, and to ensure that information will not be lost if power failure occurs.

Application

Standard integrated circuits can also be obtained to provide serial and parallel interfaces, and in an embedded system these are usually connected to sensors and transducers rather than to conventional computer terminals. A complete, dedicated microcomputer might then have the structure shown in Fig. 5.6 (note that since the ROM can only be read the data bus connection to it is uni-directional).

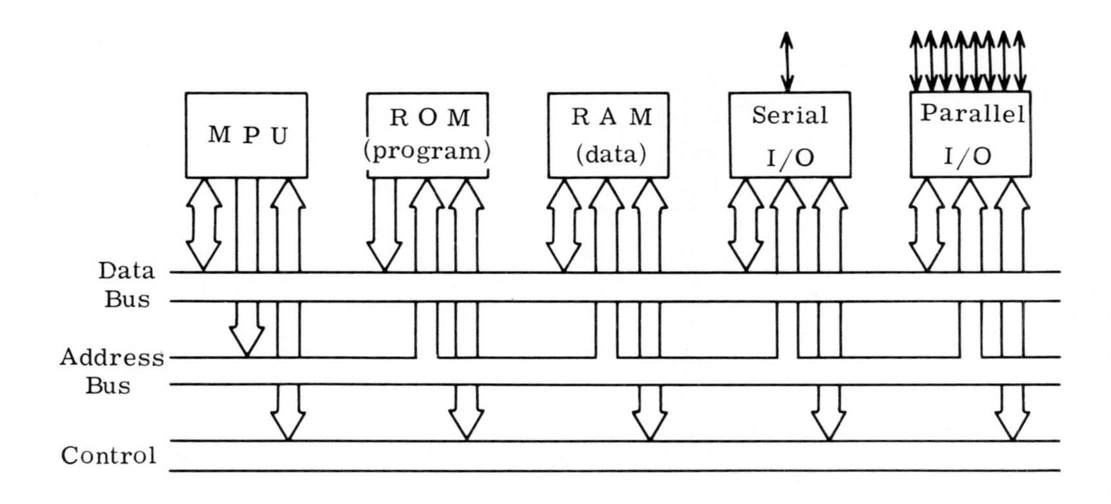

Fig. 5.6 A dedicated microcomputer

This system contains at least five integrated circuits (MPU, ROM, RAM, serial and parallel I/O) and in many applications it would be more economical to use such a microcomputer if this 'chip count' could be reduced. Therefore, for applications where the memory requirements are limited, many of the MPU manufacturers have produced single-chip microcomputers, or Microcomputer Units (MCUs). These contain limited amounts of ROM, RAM, and I/O facilities on the same chip as the MPU. A typical single-chip microcomputer would provide: 8 bit MPU, 2k bytes ROM, 128 bytes RAM, and 32 I/O lines. Undoubtedly as the number of transistors per integrated circuit continues to rise the capabilities of single-chip microcomputers will improve correspondingly.

MEMORY

Types of Memory

Semiconductor, that is transistor based, ROM and RAM are manufactured as LSI or VLSI integrated circuits, and are designed to be readily connected to the MPU buses. A number of different types of ROM and RAM are available and these are discussed below.

RAM

Semiconductor RAM is volatile, which as we know means that its data is lost if its power supply fails. Both STATIC RAM (SRAM) and DYNAMIC RAM (DRAM) are available. Data placed in static RAM remains stored as long as power is present. Data is stored in dynamic RAM as electrical charge which gradually leaks away. The data must therefore be REFRESHED, that is, read and then rewritten, every few milliseconds. Special integrated circuits are available for the control of refresh operations. Dynamic RAM has the advantage of providing more storage in a single circuit than static RAM. Currently available SRAM circuits provide up to 16k bits of store, whilst 64k bit DRAM is available and 256k bit prototypes have been developed.

ROM

ROM is non-volatile, since data is retained when the power is switched off (power-down). The data may be placed in ROM in a number of ways. Mask-programmable ROM uses a photographic mask during the fabrication of the chip to define the pattern of 0's and 1's which are to be stored. This method involves high initial expenditure and is only economically viable for volume production, when its low unit cost makes it worthwhile. Programmable ROM (PROM) uses fusible-links to define the data pattern. PROM can therefore be programmed once only, using a special circuit to burn out links as required. Erasable PROM (EPROM) can be reprogrammed many times. Programming consists of using a special circuit to produce voltage pulses which cause the data to be stored as trapped electrical charge. This data can be considered to be permanently stored since the MPU cannot overwrite it, and the charge would take decades to drain away. However, erasure of the data can be carried out by exposing the chip to a strong ultra-violet light source for a few minutes. The packages containing EPROM circuits therefore have quartz windows over the wafers in order to allow erasure to take place.

Electrically Alterable ROM (EAROM) can be both written to and erased by means of special pulses. The unit cost of EPROM and EAROM, especially EAROM, is high but their re-usability makes them suitable for use in prototypes and in low volume production. EAROM is sometimes also used as a non-volatile read/write store, although it is slow to write compared to standard RAM, and is therefore sometimes refererd to as Read Mainly Memory.

DEVELOPMENT OF A MICROPROCESSOR-BASED SYSTEM

General

The development of microprocessor-based systems splits into the usual two areas, hardware and software. Hardware development consists mainly of the selection of integrated circuits and their interconnection. Because of the bus structures supported by MPUs much of the hardware can consist of standard circuits, and so the hardware design effort can be estimated accurately, and may often be small. However the software consists of the specific programs required for the particular application. It is this split between standard hardware and specialised software which makes the microcomputer so useful, since the same, or very similar, hardware may be used for a variety of applications. Moreover a change in the way a system is required to operate can often be accommodated merely by modifying the software. Since the hardware is relatively cheap and easy to interconnect the mistake is often made of expecting the complete system development cost to be low. Very often, though, this is not the case, because of the high cost of developing or modifying software.

A major obstacle to developing the software, and sometimes also the hardware, is, as we have already seen in earlier chapters, obtaining a sufficiently accurate definition of how the system must behave. Because digital systems and microprocessors are logical devices they behave exactly as programmed, so the definition must be precise and must describe how the system is to behave in all circumstances. In the case of a system which is to perform a complicated task, this stage of the development process will be a major one.

After the system requirement has been defined software and hardware development can begin. The software may be written in either a high or low level programming language.

Hardware

Complete hardware development can be carried out using a Microprocessor Development System as a host system but often standard commercially available hardware may be used, at least at the prototype stage. However the ratio between software and hardware development effort is currently often three to one, and is likely to approach ten to one in the future as more and more powerful standard hardware appears. Consequently it is normally most important to ensure that powerful software development aids are available, and that the software development task is made as straightforward as possible. This recalls the earlier point of ensuring that a complete and correct system operation definition is available before development is undertaken, otherwise very expensive and time-consuming software redevelopment may be necessary before the system will be acceptable to the final user.

Software

If the software is to be developed in a high level language a relatively powerful computer or microcomputer is needed as a host in order to be able to run the compiler program which performs the translation. The resultant machine code can then be loaded into the memory of the target microprocessor system under development.

Figure 5.7 shows a typical development system, comprising microprocessor, 64k bytes of RAM, keyboard, visual display unit (VDU), floppy disc bulk program and data storage, and a unit for programming EPROMs. The cables and boxes to the right of the main unit enable it to be coupled to a prototype system in order to monitor and control the prototype's performance.

Fig. 5.7 A development system

CHOOSING A MICROPROCESSOR

Requirements

It is clear from the preceding section that the starting point for deciding which microprocessor to use for a particular application must be the definition of system requirements. Since there is always a danger that this definition will be inaccurate, or that additional system features may be required later, it is most important to select a microprocessor which is easily capable of handling the task as defined. If, for instance, an MPU is chosen which is only just capable of meeting the speed requirements of the system then any additional task arising later may mean that a completely new design with a different processor must be undertaken, rather than being able merely to modify the existing equipment.

Role

It is possible to make an arbitrary distinction between control and data processing applications, but in practice many systems lie between these two extremes. At the control end of the spectrum speed of operation may be critical; as may the availability of suitable I/O devices compatible with the MPU. For data processing applications the MPU word length may be an important factor in determining the speed of system operation, as may be the provision or otherwise of built-in multiplication and division operations. In addition, the area of application will determine whether initial manufacturing cost is a prime consideration, and whether low power consumption is an important attribute, or whether operation in a hostile environment is required.

An 8 bit microprocessor will often be sufficient for control applications, whilst one or more 16 bit MPUs may be needed in order to perform a complex data processing task. If cost and size are important factors then the use of a single-chip microcomputer will be worth consideration. Low power consumption can be achieved by operating the equipment intermittently or by the use of CMOS (Complementary Metal Oxide Semiconductor) integrated circuits for the MPU, memory, etc., since these use very much less power than the more commonly used ICs. The need to operate in a hostile environment may mean that only MPUs designed to meet military specifications would be acceptable, and not all MPUs are available in military tested forms.

Support

Regardless of how well a microprocessor appears to meet the system requirements, it will, in practice, be almost totally useless without effective software and hardware support to aid the system development. This means that hardware development systems, compatible memory and I/O circuits, and high and low level software support should be readily available. Also it is important that there should be a reasonable likelihood of the MPU continuing in production during the lifetime of the equipment in which it is to be used.

If a new microprocessor appears and has some measure of hardware and software compatibility with existing devices whose use has already been mastered by a design team then it will offer a distinct advantage in terms of ease of use, and therefore cost, over other new but unfamiliar devices. Therefore once an MPU has been chosen for one particular application it is often sensible to try to standardise on the use of a family of compatible devices.

RELIABILITY

Hardware

Microprocessor systems can be considered in terms of hardware and software reliability. The failure rate against operating time for the integrated circuit hardware is usually found to give a 'bathtub' curve as shown in Fig. 5.8. There is an initially high failure rate during what is known as the 'infant mortality' period. The failure rate then remains low for a relatively long time, rising eventually as after years or decades ageing effects begin to occur. The dangers of the infant mortality period can be overcome by 'burning-in' the circuits. It involves running them in a test system, for a longer time and then discarding the failed circuits. Only the proven ones are retained. MPU manufacturers normally only perform full burning-in to special order, or for military specification devices. The life of the hardware can also be adversely affected by extreme environmental conditions, and in particular exposure to ionising radiation.

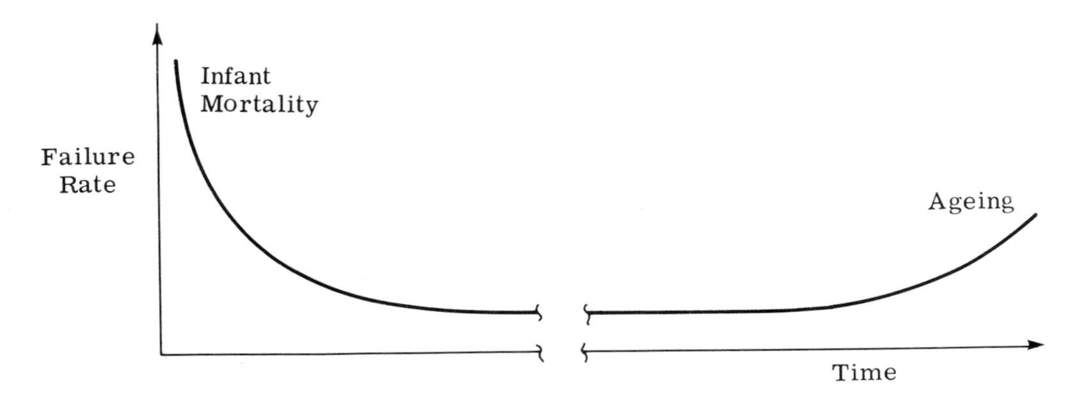

Fig. 5.8 Failure rates

Software

Software reliability is a more difficult concept to quantify, since corruption of data or program is really a failure of the hardware storing it. System failure caused solely by the software is due to the fact that some error was built in to the program when it was developed. Unfortunately such errors have a habit of only appearing after equipment has been in service for some time as mentioned before.

APPLICATIONS

Range of Applications

There is a wide range of possible applications of microprocessors, from data processing where the MPU forms the basis of a low cost computer, to dedicated control, where it is an integral part of a control loop. In between these extremes there are many applications employing a mixture of data processing and control. Microprocessor-based systems are already in production for a large number of these applications, and any attempt to list them all would result in several pages of tables, which would be out of date and incomplete before this book went to press. Some military applications of microprocessors are dealt with in Chapter 7, but since many commercial applications carry implications for future military usage a selection of these is included here.

Personal Computers

An application very much concerned with data processing is the home or personal computer. The appearance of these microprocessor-based very low cost computers has resulted in the emergence of a completely new market for electronic equipment. Not only do many small businesses make use of personal computers for stock control, accounting, order processing, and so on, but also there has been a rapid growth of interest in computing among the general public, with many computer hobbyists buying their own personal computers. These machines can also be used in scientific research for control of apparatus, data logging, and data processing, in fact they can be used anywhere a larger, conventional computer could have been used, provided only that their slowness and limited memory pose no problems in the particular application.

Fig. 5.9 A personal computer

A popular personal computer is shown in Fig. 5.9. This machine uses an 8 bit MPU, and can be purchased with up to 32k bytes of RAM. It has an integral keyboard and display, although many other personal computers are designed to make use of a domestic television as display, and bulk program and data storage can be on cassette tape or optional floppy discs. A printer may also be connected to it if a permanent copy of results, or printed documents are required. Whilst a computer hobbyist is likely to be keen to write his own programs, it is vital if a personal computer is to be of real use in business applications that effective software packages be available to provide the functions required. To this end a new industry is emerging, consisting mainly of small, specialised firms, to produce this standard software.

Word Processing

One blossoming business application that can be undertaken by personal computers, but for which purpose-designed and programmed systems are also available, is word processing. Instead of a typewriter being used to produce a document the computer keyboard is used to enter the details, and the text appears on the visual display and is stored in the memory. The typist may readily edit the text, deleting or inserting sections, and the processor automatically adjusts the layout and pagination to suit the required format. The final document is produced on a printer attached to the word processor. The power of this approach lies in the fact that if many similar documents are required, or if a number of standard paragraphs are to be used, these can be called up from the memory and the typist has only to enter the details specific to that particular document. For instance if a circular letter were required then the typist need only enter the letter once, omitting the specific details of recipient's name and address, then enter or call from memory a list of recipients, whereupon the word processor will produce the required number of letters, each apparently individually typed, addressed correctly to each recipient. Word processing and the use of word-processors communicating directly with each other removes the need to send physical documents through the post, and here are the first stages in the appearance of the 'electronic office' in which many of the tasks of document preparation, storage and transmission will be carried out automatically, under the control of microcomputers.

Robots

One of the most widely publicised control applications of microprocessors is in industrial robots. The movement of a robot arm is controlled by a microcomputer, and if a different task is to be performed then it is only necessary to enter the corresponding program. Programming such a device could be very difficult, but a simple method has been devised. An operator initially guides the arm through the required movements and a record of them is stored in the processor's memory; the robot is learning the task. Subsequently the machine can be left to operate unattended, controlled now by the stored information. If a robot is to perform a number of different tasks then the various programs will be held in a backing memory, but if only one task is to be performed then an EPROM will often be used to store that program.

Vehicles

Microprocessors are in use in control and display systems in motor cars and aircraft. One application in the motor car is the control of engine ignition timing. The memory holds a look-up table of advance/retard values for varying engine speed and inlet manifold pressure; sensors read in these terms and the MPU obtains an advance/retard value from the table and produces a control pulse at the correct time for ignition. This type of system can be much more accurately matched to the engine's requirements than the mechanical method it replaces; and this means that its use can improve the fuel consumption and performance of an existing engine. Microprocessors are also being used to control fuel injection, gear selection, braking, and suspension, and to produce improved instrument displays. The harsh environmental conditions encountered under the bonnet of a motor car have meant that although the advantages of using microprocessors have been appreciated for some time, it is only relatively recently that reliable systems have appeared in mass-production vehicles.

Consumer Products

Relatively low cost consumer products such as sewing and washing machines now make use of microcomputers. Since these products are produced in large numbers the development cost is not usually as important as low parts count and component cost, which means that these are ideal applications for single-chip microcomputers. The area of ROM holds the program controlling the machine's operation, and the I/O connections enable controls and transducers to be connected. For instance in the case of a washing machine there will be switches to enable the user to select the wash programme, a transducer to measure water temperature, and circuits to enable the MCU to control the water valves, motor and pump. If new wash routines are required later in the production life of the machine then these can be accommodated by altering the program used in the MCUs for these later machines. The cost of this compares favourably with the redesign of the complicated electro-mechanical controllers used in earlier generations of machines. The replacement of these controllers by microcomputers has also resulted in an improvement in machine reliability.

Industrial Plant Control

Microcomputers are also coming to be used increasingly for the control of industrial plant such as chemical processing systems and production lines. In these cases the number of microcomputers used in any one application is small, so low unit cost of the hardware is not usually a dominant factor. The development cost is usually more important since this may only be amortised over a very small number of units, sometimes only one. In the past minicomputers were sometimes used for this type of application. They were easy to use, but relatively expensive and large, and required housing in a protected environment. In theory personal computers could be used, but they would be unlikely to survive long on a factory floor. Especially rugged systems are available which may be used in most environments, but an alternative in all but the most severe cases is to use standard microcomputers built on a single printed circuit board. A variety are available, but broadly speaking they provide similar facilities and are intended for housing in industrial equipment racks or cases. One board is shown in Fig. 5.10; it provides an 8 bit microprocessor, 16k bytes of RAM, up to 4k bytes of ROM or EPROM, serial and parallel I/O ports, a timer, and interface circuits to allow it to be easily connected to other boards to provide more memory or I/O facilities if required. Development of the software for use with such standard boards still usually requires access to the facilities of a microprocessor development system.

Fig. 5.10 A microcomputer on a single board

Instruments

Another application of microprocessors is in an 'intelligent' theodolite. This device uses a laser to measure the distance from the theodolite to the object being surveyed. The microcomputer pulses the laser several times and decides the distance by selecting the value most commonly given by the laser. Then, given its own orientation which is measured by tilt sensors and rotary encoders, the microcomputer can calculate the co-ordinates of the object. The results may either be recorded manually or fed straight to an automatic data logging system of some kind.

Science

There are many applications of microprocessors in such diverse areas as medicine and horticulture. Patient monitors, and such equipments as scanners, frequently use microcomputers to provide control and data processing functions. In horticulture work is underway investigating the application of personal computers and dedicated microcomputers to problems such as the control of growing conditions in a greenhouse. Temperature, humidity, soil moisture content, and light conditions are measured by sensors connected to a microcomputer which then controls ventilation, heating, watering and lighting in order to ensure that the required growing conditions are always present.

Games

Many games and entertainment devices use microprocessors as their controllers. Educational systems are also being produced, where the entertainment value of using them encourages the acquisition of some skill or knowledge. An example of a system of this type is shown in Fig. 5.11.

Fig. 5.11 Speak and spell

The unit contains a single-chip microcomputer, a special speech output chip, and ROM which contains the information required to define the sounds to be produced. The device is intended to help children to learn to spell; it speaks a word, then asks the child to enter the correct spelling. If the spelling is correct the machine moves on to another word, if not it gives the child a second attempt before itself giving the correct spelling. After every ten words the child's score of correct spellings is given, and as the child progresses he or she can select the level of difficulty of the words to be spelt.

Future Applications

It can probably be said more accurately about microcomputers than any other modern development, that their applications are only bounded by the limits of human imagination. During the next generation they will be involved in most home and work environments.

SELF TEST QUESTIONS

QUESTION 1 To what part of a traditional computer does a microprocessor correspond?

Answer ....

QUESTION 2 What is a silicon chip?

Answer ....

....

QUESTION 3 Identify the main factors determining integrated circuit package size.

Answer ....

....

QUESTION 4 If a microprocessor has a 20 bit address bus how many distinct locations can it access directly?

Answer ....

....

QUESTION 5 Why are bit-slice microprocessors sometimes used in preference to ordinary microprocessors?

Answer ....

....

QUESTION 6 Describe the benefits and limitations deriving from the use of a single-chip microcomputer.

Answer ....

....

....

QUESTION 7 Indicate why it is often the software aspects of microprocessor system development which consume the majority of time and money.

Answer ....

....

QUESTION 8 Why are some integrated circuits 'burnt-in'?

Answer ....

....

QUESTION 9 Identify current commercial microprocessor applications, and suggest possible future ones.

Answer ....

....

....

ANSWERS ON PAGE 201

6. Security

INTRODUCTION

Problem

The problem of security of a computer and the data or information it may contain has received much publicity since it became apparent that computers can play a major role in processing and storing secret, private or proprietary information. They can, therefore, possess tremendous capability for effective dissemination, and thus sharing, of useful facts. There is the added problem that whereas a single item of information may, on its own, be almost irrelevant and worthless the aggregate of several small items of information may present something far more useful in terms of potential value to an enemy, whether that enemy is a business competitor, a military opponent or even somebody with straightforward criminal intent.

The subject of security is therefore worthy of a separate chapter in this book, where the problems and some suggestions for solutions are discussed. The main task is to consider the nature of security and to do this it is necessary to go back a step to the requirement for it.

Privacy

In a computer context privacy can be considered to be the legal and moral requirements to protect data from unauthorised access and dissemination. These requirements can range from industrial and commercial confidences to military secrets or the invasion of personal information so much at risk in modern society. Privacy issues are decided at a legislative level; the decisions are normally long and drawn out due to the many lengthy iterative stages, and of course take no account of the financial and technical penalties which will have to be imposed upon computer systems. These penalties are difficult to quantify because they have a global impact. For example, at first sight it would appear senseless to guard against publicising the cost of a computer system or component; the cost is, however, proportional to complexity and complexity may well be an indication of potential, and thus cost could mean more than pounds or dollars. What, then, of the measures necessary to achieve and enforce privacy?

Security

Security is considered here as the managerial procedures and technological safeguards applied to computer hardware, software and data to assure against either accidental or deliberate unauthorised access to, and dissemination of any data held in a computer system. The range of measures involved in the security plan is large and the measures are often interrelated and thus confused; to describe them they are divided into two sequential stages, the first concerned with denying access to the system and the second with protection during the actual operation of the computer.

DENIAL OF ACCESS TO THE COMPUTER

Operational Considerations

Based on the legislation and security requirements some form and level of operational security is established. This will allow the management to exercise control. Data classification, system configuration and levels of control will be established. In other words the ground rules for authorisation will be defined and promulgated. It is not an easy task for there will be a balance to be considered; this will include the impact of the constraints both on the organisation, the efficiency and the cost.

The requirement for access must be identified at the outset. It will be necessary to determine what data may need to be accessed, then who will need to be involved, at which terminals do those people operate and when will they need access. Finally it will be essential to establish what operations those people will perform on the data and the reason why. Defining such a requirement for a small system consisting of one or two CPUs and five or six terminals should prove relatively simple but it is easy to see that, for example, a battlefield command and control system with many users and many terminals could prove to be a large task.

Denial of individual access can be used to limit the number of people who are allowed access to the system. It is possible to provide some flexibility by perhaps specifying that, say, all dialogue with the computer must be confined to one of a very small team of operators who do have direct access, or allowing any operator to prepare work and then to screen the individual before the actual processing is undertaken by the computer.

Ease of operation is often left out of the considerations with the result that, in practice, the security system does not run well and, in certain cases, does not work. It must be recognised from the outset that security will bring extra work, cost, and inconvenience; it must be explained to the operator as a very necessary burden; he must be encouraged to accept security as part of his profession, whether it be the need to remember a password or the apparent waste of time taken to check in and out of the computer area.

Recovery procedures for computer systems are often based on duplication of data in back-up files. Such duplication will of course increase the security task but it is possible in certain cases to use special software programs designed to test reliability to also carry out some degree of security check.

Selection of personnel to work in a secure computer environment is a similar process to that used for personnel working in other secure areas: the principles of division and layering of responsibilities and frequent checks are employed. Division of responsibilities should be essentially sequential to ensure that no one programmer completes an entire software task on his own. It can also be effected by separating responsibilities between various operators in order that no single person is given the opportunity to work with classified data on his own for an unacceptably risky length of time.

Cost and economic considerations will need to be taken into account. The balance will be between the best security available and the value of the information concerned. In a military environment the information will be critical if it is held in one computer store only but, by distribution of the information to all those cells which might need it, a level of security against loss has been built in; correspondingly, of course, the fact that the information is in more than one place has increased its vulnerability.

Physical Security

Once the operational considerations have been made the actual measures to provide physical security and to deny access to the computer must be implemented. In military operational systems normal measures including sentries and personal identification are employed; these can, of course, be supplemented by extra perimeter protection where necessary and the introduction of limited/named access, again using the normal sentry procedures. In static locations it may be sufficient to employ 'password' systems involving coded pushbutton door locks or similar devices; but these are only likely to be acceptable in areas already subject to some other access restriction like a permanent patrol. Thus physical security to prevent a person or persons from reaching a computer can be achieved; the problem raised by the computer system emitting electro-magnetic radiation which can be detected and interpreted by other machines is rather more difficult. In military operational systems it is possible to select components and architectures which are proof against this type of emission but such solutions are expensive and commercial equipment does not normally have similar protection.

Having allowed persons to pass through the physical security barrier whether by design, as in the case of an authorised operator, or because the physical system is infiltrated, the following levels of security are applicable during the actual running of the computer system.

DATA PROTECTION MECHANISMS

Data protection mechanisms can be divided into three areas, hardware, software and the database itself.

Hardware Security

Hardware consists of CPUs, store or memory, and peripheral devices. In early systems no hardware security was employed but it is now possible to design such features in from the concept stage providing the requirement is clear, for it will prove expensive to incorporate them late in development. Hardware security mechanisms vary from simple to highly sophisticated, but will not normally be included in hardware unless specifically requested.

The hardware components are the containers for the data which has to be protected and hardware security is concerned with making the containers secure. One of the most difficult problems facing the designer of such secure devices is at what level to include them. For instance data is held in groups in bytes, words and files so the aggregates can be small or large and to complicate the situation the data will be moving around the system in concurrent programs. Obviously the level at which the security devices are incorporated will depend to a great extent on the balance between cost and technology and also the balance between cost and the value of the data. The problem is to decide at what stage will it prove non cost-effective to buy more protection. There are very few guidelines available but the advent of the microprocessor has meant that possible solutions should become cheaper.

In simple terms hardware protection devices will restrict the operator's access to data and files. There are several modes of operation:

1. Read-only, where the operator is allowed to read data from files but cannot alter it.

2. Execute-only, where the operator can run a program but cannot dissect it to discover its logic.

3. Non-accessible, so that any attempt to access the data will result in an alarm or interrupt.

4. Recording access, whereby a record of the identity of the operator, the date/time, and the reference of the data accessed is made, thus discouraging, rather than preventing, access; but at least it provides an accurate record of access.

The hardware protection device is used to verify that the store or memory locations which will be accessed in order to carry out instructions requested by the operator are in fact not subject to any pre-arranged access restrictions. Unless the device authorises use of the locations the CPU will not be able to complete its task of interpretation of the instructions. At first sight this scheme appears to be straightforward but it must be remembered that multiprogramming allows several programs to run concurrently in a system; more than one may attempt to access the same store or memory location simultaneously and it is possible that each will have a different access limitation. Such eventualities can be provided for by setting up a hierarchy of overriding programs to which the protection device can refer.

Software Security

Systems software, the operating system, manages and controls the hardware resources and is thus involved deeply with security where its roles are to identify the operator and then to supervise the access procedure. The main principles are:

1. Logging-in. Whatever the type of computer system it is necessary for operators to log-in at the start of operations and to log-out at the end; this provides a suitable mechanism for the operator to be authenticated by reference to personal details already held in the database. Once such authentification has been carried out it may be necessary to repeat the process at intervals to ensure that the same operator is at work. This may have particular relevance in military operational applications where there is a risk of capture of terminals.

2. Access. The completion of the authentification process will enable the operating system to itemise all the programs and data to which this particular operator is allowed access in order that, from then on, each instruction can be considered from that point of view. The itemisation can be best understood by considering it conceptually as a matrix of operators against programs.

3. Isolation. The situation described so far, of an operating system managing and controlling all the hardware in a multiprogramming environment, identifying operators, and controlling access to data is clearly a complicated one and it is possible for security to be breached within the system. A policy of isolation is normally therefore employed in which only one single operating system is used to carry out all the monitoring of a set of programs. Several operating systems can exist in the same computer but again each will have control of only its own associated programs. This concept is known as the 'virtual machine', for each group of programs believes it is the only group in existence. This is an important isolation technique because security breaches will not easily leak from one group of programs to another and two different groups of programs can therefore be afforded different security levels. However it must be emphasised that this latter benefit is applicable only to programs. It is not always considered practicable to store data of different security gradings in the same computer due to the risk of unapproved access and the fact that it is impossible to erase data completely from a store even after much overwriting.

The systems software, or operating system, which carries out all the work to achieve software security must itself be secure. The methods used to check that the design meets the program specification are called VERIFICATION. This process ensures that the programs meet the intent of the specification and various techniques exist to prove the programs but these techniques have limitations concerning their ability to ensure security. In particular the operating system may be complex and as verification processes are necessarily long it may not be cost effective or indeed possible to test the whole operating system in this way. In these circumstances it is possible to identify those parts of the operating system which control the areas of security risk and to assure the security of these parts, which are called KERNELS. They will normally cover store/memory, logging, and access control.

However carefully the kernel is identified, and the software is verified, it is reasonable to expect some areas of the kernel to be less than totally secure. A method of reducing the chances of a security breach even further is to adopt a process of PENETRATION TESTS which will hopefully identify weak spots in the software. If a sufficiently comprehensive and intensive series of such penetration tests is employed on an operating system kernel during development it will allow security certification of that product. It will then provide a measure of 'security confidence'. It must be stressed, however, that absolute security confidence is impossible to provide because the series of penetration tests cannot identify every possible weak spot.

Database Security

In considering the security of the data in a database it is important to recognise the two main ways in which data is treated. Data processing is the traditional concept which treats data simply as a collection of values. Programs are written which know the attributes of the data and which process the values in accordance with their type. Processing of the values can easily provide useful information, but on-line access is not possible and neither is interactive use of the data. The computer staff must run the programs for the user and return results to him. If the results are incomplete, there is no way of knowing why or whether the user is authorised to have the missing data or not.

Database management is the more modern concept. It involves the stored values of the database. It also gives the users a uniform view of the database necessary to support on-line access, dynamic update of the database and multi-user interaction. Thus there must be a system of recording the relationship between pieces of data and how those pieces of data are used. It is a well structured way of keeping track of data.

A number of factors must be considered when considering access to a database and the rules laid down within the computer system are based on the following types of qualification of the information:

1. Event-sensitive, where, for instance, it is possible to restrict access to a period between two previously agreed times.

2. Value-sensitive, where the decision is based on current values of data; for instance it would be possible to deny access to any data of a 'value' of £1000

or more. A number of database management systems have been designed to provide such protection.

3. Pattern-sensitive, where prescribed usage of data is involved. In this case it would be possible for a program to be used to process data but where the program itself could not be examined.

4. History-sensitive, where protection is provided against an operator gaining information to which he is not entitled by inference or reference to another file. As an example there may be two files, the first listing salary against rank and another listing rank against name. Free access to both files would enable a list of names against salary to be composed thus defeating the aim of separating the two original files for privacy reasons.

Once the database management system has granted access to the database there remains the problem of accessing the necessary data. In an ideal situation the database management system will access only the data involved in the agreed request but in practice it may be necessary for a search to be made through, say, data classified at a variety of levels to find the piece of data in the request; thus the true access is to all the data searched. Such problems are termed PASS-THROUGH and the optimum software design will of course strive to eliminate pass-through problems totally.

One obvious solution is to compartmentalise data into groups warranting similar protection requirements and most military software will be designed in this fashion in view of the multilevel security involved. Another solution is to develop the concept of ownership of data to allow protection attributes to be passed to different users. In this event a user can deny and grant access to other users of data areas if he is the 'owner' of that data. In most cases the 'owner' is the creator of the data and he is able to place any security attribute on that data including permission for another user to be a co-owner. This solution is particularly relevant in the design of battlefield command and control systems where there may be an intricate matrix of owners and levels of security, and where it may be vital that only the owner or creator of data should be allowed to change that data. It can be seen that data on minefield locations should only be altered by the person who actually changes the configuration of that minefield.

An integrated approach to hardware, software, and database security is quite obviously necessary to avoid cost and to ensure total security; but such integrated approaches are as yet uncommon. However advances in processor technology, such as microprocessors, and memory technology are encouraging and a good deal of effort is being spent on the problem.

TRANSMISSION OF DATA

A computer system in anything other than the smallest configuration consists of a network of processors, terminals and storage devices. Such networks require communications between the components and cryptographic techniques are used to encode data to hide its content during transmission. Several systems for such encryption exist already and many more are under development. The main decisions affecting the design are the degree of security required and whether such a system should be based on a hardware component, a software algorithm, or a combination of the two.

The process involves the data, a cryptographic key and a computer operation. By performing the operation, data and key digits data is encoded at the transmitting cell and decoded on arrival at the receiving cell. Keys will differ in size according to their function. In a message oriented system where the messages and packages of data are, by design, short a reasonably short key will suffice. In information oriented systems, where messages and packages of data are longer the key will be necessarily longer. This latter effect can, in fact, be achieved by the use of multiple short keys. Frequent changes of keys increase security but will necessitate mechanisms to produce random patterns of keys.

Cryptographic Systems

It must be assumed that components of the computer system involved may fall into the hands of those wishing to have access to the data being transmitted. This means the only protection available for that data is the security of the frequently changing keys. Moreover it must be assumed that any enemy capturing an actual key can, given time and the necessary equipment, decode the key. The cryptographic designer must therefore take into account the threat in all its forms, including the enemy's desire or ability to spend sufficient funds to field decoding systems.

Various national and international agreements and rules have been developed. Designs of cryptographic systems are becoming plentiful and perhaps the least expensive are software based, but this solution is only now becoming acceptable in terms of confidence. For instance the US Data Encryption Standard intended for commercial systems uses a 56 bit main key providing 2^{56} (or approximately 7×10^{16}) ways of encoding each 64 bit data word that is transmitted. This is believed to ensure that even with the aid of the largest computers available it would take years to break the code in any message. Yet of course some doubt, although very slight, still remains that there might have been a flaw in the encoding algorithm selected, which could be found and used to enable the messages to be broken very much more easily. It is very difficult to prove that such a system is free of flaws in logic. All that can be expected is a high degree of confidence that the encryption cannot be broken within the time that the data would remain of value to an enemy.

SOFTWARE INTEGRITY

Most of the problems of computer security discussed so far have been concerned with users and developed programs. In each case it has been assumed that the programs were produced honestly. This may not, of course, be the case. How can a developer guarantee the allegiance of the programmers and thus the integrity of the software? A modern technique is the employment of a Software Integrity Team, ideally on a one-to-one basis (that is one team member to each programmer) to check every piece of software for authenticity. Such a plan is

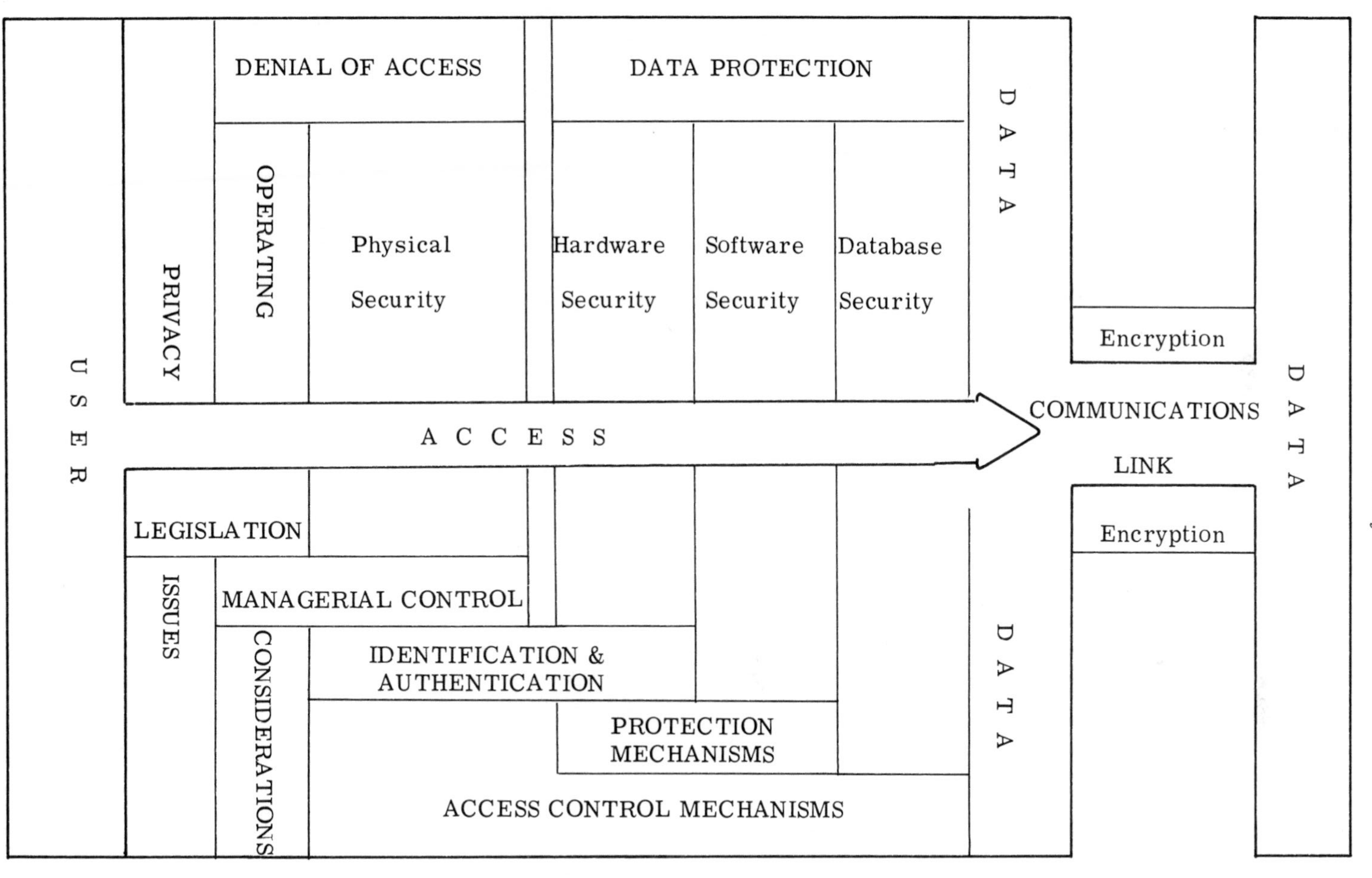

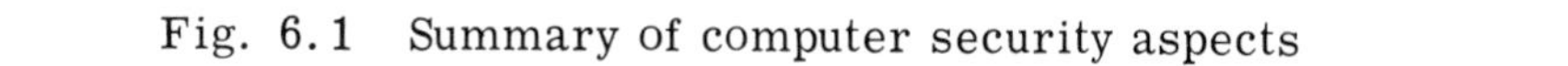
Fig. 6.1 Summary of computer security aspects

costly and cumbersome and its value must be balanced against the value of the software and the data it will process.

IMPLEMENTATION

It is worth emphasising that most people fail to understand that security measures and precautions are bound to cause inconvenience to users and that if the measures are not easy to incorporate or operate they will not be used effectively. This is a fact of life. Also the most vulnerable time for a system is during the transition period from the previous system, especially if that were a manual system. This vulnerability to security violations can be attributed to carelessness, ignorance, or the fact that the security measures built into the new system are not fully proven before implementing and are thus possibly inadequate. Therefore it is important that extra emphasis be placed on security during the transition phase.

SUMMARY

The security of information held in a computer is protected by physical means, administrative procedures and computer software and hardware. These types of security measure are interdependent and can, in conjunction, make the risk of information loss very small; but none is adequate in isolation. Especially, all-purpose software 'systems' or 'packages' are often neither safe nor appropriate. A complete security environment must be designed and set up to meet the special needs of particular systems. The speed and power of the computer creates problems for which there are no easy answers. Only the systematic application of established security and administrative procedures of proven effectiveness in other fields, the appropriate use of the tools provided by the computer itself and conscientious efforts by everybody associated with the system, can keep the data effective and secure. Figure 6.1 shows a summary of computer security aspects.

SELF TEST QUESTIONS

QUESTION 1 How can access to the computer be denied?

Answer ....

....

....

....

....

QUESTION 2 What are the basic functions of a hardware protection device?

Answer ....

....

....

QUESTION 3 What is the purpose of software verification?

Answer ....

....

QUESTION 4 Suggest some measures to protect the data in a battlefield ADP system.

Answer ....

....

....

....

ANSWERS ON PAGE 202

7.
Military Applications

INTRODUCTION

Computers are now an essential part of practically every military system; the armed forces of the world use computers on a very large scale and there are few aspects of military life that do not depend upon their efficient use. In this chapter we will look at a cross section of their applications.

MILITARY ADMINISTRATIVE COMPUTERS

Military administrative ADP systems are not generally considered as battlefield systems but it is important that they should be mentioned here in order that their role in the systems approach to computing should be highlighted. They can be considered in two groups; those which have everyday roles in common with civilian computers and those with specific military applications.

Normal Administrative Roles

A computer bureau is based on a centralised computer which offers a service to a variety of possibly random users who may either bring their work to the bureau or, if they are going to be regular users requiring a reasonable response time, may install a direct communication link with the bureau. Such systems are common in the commercial world and save organisations with only small quantities of work or little technical expertise from having to purchase and install their own dedicated systems. Military bureaus are often used for statistical and movement planning tasks in headquarters.

Computers are also in common use in the production, engineering and maintenance of military equipment. For example it may be considered necessary to report repair, workshop and reliability data from Army workshops to a central computer. The computer could analyse the data and produce advice to assist in the efficient management of equipment and repair resources and could provide information on equipment reliability and requirements for spares.

Financial management in army departments can be aided by the use of computers, as can the payments and recoveries of contracts in the equipment/procurement field. Project managers require timely, accurate financial information of many types in order to exercise effective control. The same computers can be used to provide information retrieval systems with such tasks as literature searches in support of military projects and contracts.

Most armies today use computers to handle pay accounts and in many cases it has been found sensible to include personnel documentation, records and medical information in the same database.

Supply and stock control systems, too, are fairly common. There is normally a tiered structure of processors starting at depots and gradually fanning out in a distributed form to smaller depots and operational formations, possibly in other countries, using a range of communications media from line to satellite. The extent to which the processing facilities are deployed in the field varies from one nation to another but it can be easily appreciated that a data link from divisional level stores units back to the rear base depots is at least, highly desirable. The principle in such stock control is that stock holdings are automatically controlled by the next level up and that cost savings should be made by economising on surplus stock levels; at the same time it is essential that the right stock should be held at the right depot at the necessary time.

Specific Military Uses

Whilst many military administrative systems use commercial hardware and relatively common software as can be seen above, others need a good deal of enhancement before they can be effective. A good example is training equipment. There is a requirement, within any army using ADP equipment, to have a sufficiently large number of personnel trained in the operation of the equipment, maintenance of the equipment, design and implementation of future systems and with appreciation of the application of ADP. Most technically advanced armies will thus have a number of training systems, often deployed at Arms Schools, providing a variety of training facilities. The range of requirement is wide, from individual components and full system test beds used to teach fault finding and maintenance on both components and system at engineering establishments, to specific-to-Arm systems, often using operational hardware and specially designed training software to train operators and commanders. In both cases the software can be very costly and it is essential that the user thinks ahead to his requirement in good time.

Another example is operational analysis. Most nations are involved in operational analysis to some extent. Many employ large computer systems on which to process the many options which they develop for various aspects and phases of operations. The continual comparison of possible alternative procedures, equipments and strategies against pre-defined models is critical in development of those areas. Without ADP, results from such tasks would be almost valueless.

An important field is that of war-gaming, the modelling of imaginary operations. Sophisticated computer aids have been developed to hold battlefield organisations and to calculate loss rates in accordance with the various moves in the game. The degree of definition in the model and of the interactive responses are limited by cost in that they are software based developments and therefore expensive; so too are soldiers, fuel, and training areas so a sensible balance must be sought in order that computer aided training devices can be used to supplement live training cost effectively.

Systems Approach

It is not difficult to trace user threads which are common to administrative and operational equipments. There is the need for the logistic function to be based on one system from forward stores units to base depots; the need for operators and maintainers to be trained on operational equipment as far as possible; and the need for operational analysis to be based on what the user can actually do in practice. It is therefore essential that the two areas of military ADP equipment should not be allowed to evolve in isolation but that they should be mutually dependent.

COMMAND, CONTROL, AND COMMUNICATIONS (C^3)

The more obvious military applications of computers, in weapon systems and command and control systems, are expanding rapidly and few operational systems do not now include at least one computer controller. Many include several computers linked together by sophisticated communication systems allowing data, voice and information to be passed throughout all levels of command.

C^3 has always been a vital aspect of military operations, but only recently has it covered such an extensive amount of sophisticated technology. C^3 now needs much greater consideration and emphasis to keep pace with the development of both the weapons systems and forces to be controlled, not to mention those of the enemy.

Recent technological advances have had a significant impact in such areas as missiles, sensors, and satellites for communications, surveillance and navigation. Today's weapons systems have greater range, speed and accuracy; at the same time intelligence, surveillance and reconnaissance improvements are producing more information and over greatly increased distances.

The real size of the battlefield has increased. Sensor coverage and capabilities now overlap and need careful co-ordination. Consequently command and control of forces will depend to an unprecedented degree on communications and the ability to process the information, on which decisions are made, fast enough for it to be relevant. It is also important that intelligence and operations should be welded together and that their almost bureaucratic separation should be ended.

C^3 is evolutionary. New C^3 systems are not developed; they are, rather, continuous modifications of old systems. But improving a C^3 system requires the careful integration of equipment development with the projected operational task and the organisational structures and organisations involved. It must therefore be a subject for management at the very highest level and it is imperative that correct user involvement in the development process is arranged.

Command and Control (C^2)

It is worth considering the meaning of the phrase 'command and control' (C^2). A common definition is, 'the process by which a commander exercises authority and direction over his forces in the accomplishment of his task'; this emphasises the functions or processes which must go on. Considering C^2 as a process brings up the questions of what the output of that process is. The aim of an army must be to provide the means to establish or maintain control over some geographic area; thus the purpose of C^2, defined in very broad terms, is to either maintain or change the equilibrium of the tactical environment as determined by higher levels of command.

The C^2 system must therefore be designed to inform the commander of the state of his environment, help him to compare it with the objectives set by higher authority, help him to decide on a course of action and then disseminate instructions to implement those decisions.

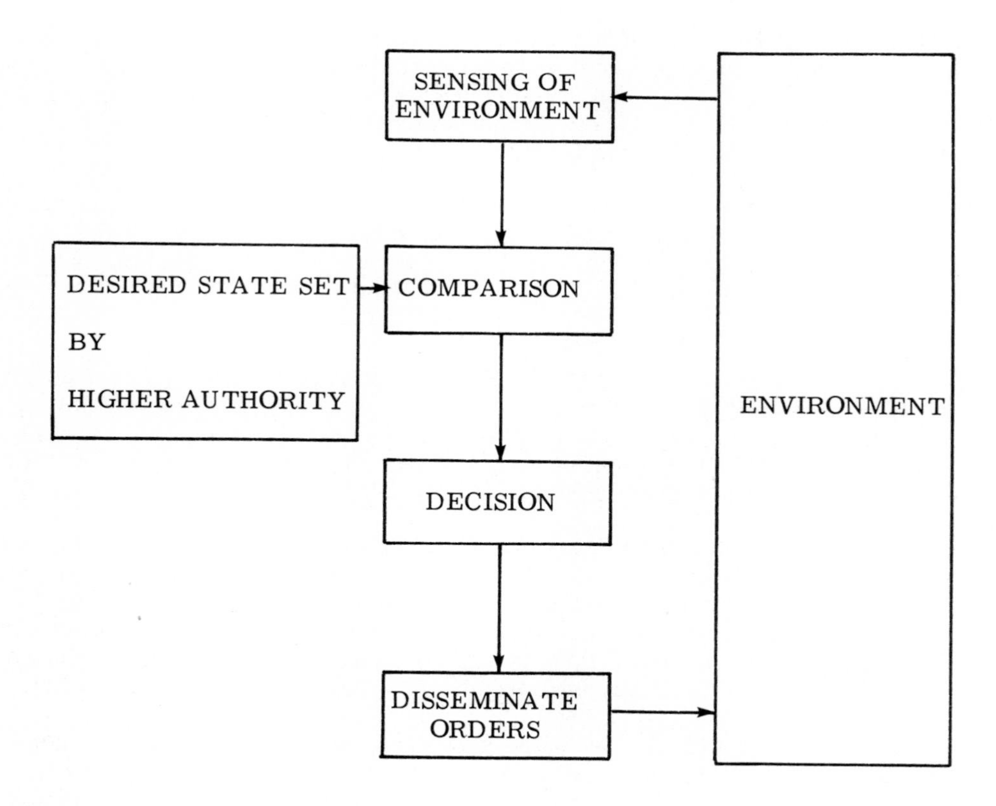

Fig. 7.1 The C^2 process

It is clearly important that 'sensing', consisting very largely of surveillance and intelligence, should be considered as part of the C^2 process. Figure 7.1 is a very simple block diagram of the process which of course could be developed to include other inputs and outputs for, as was explained in Chapter 1, every system is part of another system.

Probably the most advanced and complex example of a computerised command and control system is the US 'World Wide Military Control and Communications System' (WWMCCS) which integrates a number of other computer based communications systems into a network covering many countries around the world. We, however, will confine ourselves to the more localised army problem in a single theatre of operations.

BATTLEFIELD COMMAND AND CONTROL SYSTEMS

The complexity and intensive pace of modern warfare impose increasing demands upon commanders at all levels. Rapid decisions are required on diverse matters and immediate access to current, consolidated and relevant information is thus essential. The implementation of decisions requires a means of controlling widely dispersed resources, while the essential close integration of the operations of the various arms and services demands regular exchange of status reports covering a wide range of activities.

The common way of meeting these conditions is to establish headquarters where data and information are fed in and out of a decision making module. Management within the organisational structure of the headquarters will depend upon the pre-determined decision making levels and the nature and relative importance of the independent participating organisations. It is possible either to concentrate all the functions and tasks into one focal cell within the headquarters or to distribute them into cells which are subordinate to a main cell. However they are deployed, headquarters can be considered to consist of three functional cells, operations, logistics, and intelligence as shown in Fig. 7.2.

Each of these functions will use data and information as a basis to make a variety of decisions; the stages in each process are as shown in Fig. 7.3. This appears to be a straightforward sequence of events but in practice the information or data must be acquired, validated, stored, retrieved and disseminated; the necessary communications links must be fast, accurate and secure and they must have inherent survivability. It is in this handling of information that computers can be used to great effect.

ADP can be introduced into headquarters to support C^2 systems by providing the commander and his staff within a formation with access to a common database. It is usually effected by means of VDUs interconnected by data circuits over trunk and net radio. Such support is generally tailored to match the user's command infrastructure and procedures. First generation ADP systems based on original software development, or on the adaptation of systems intended for commercial applications, have often been characterised by extended development timescales, escalating costs and heavy demands for communications capacity.

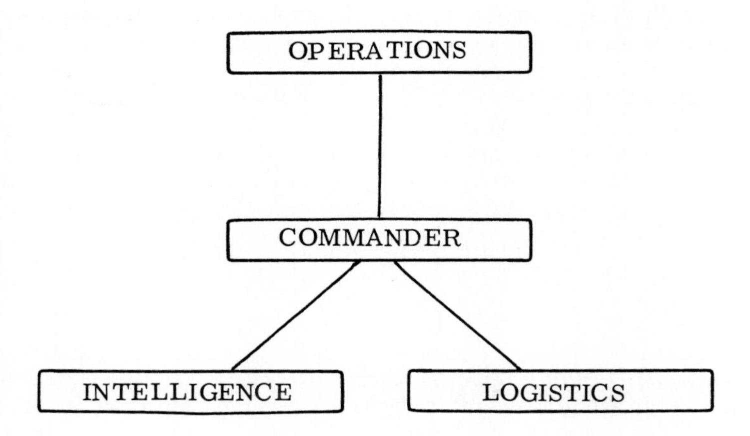

Fig. 7.2 The three cell headquarters

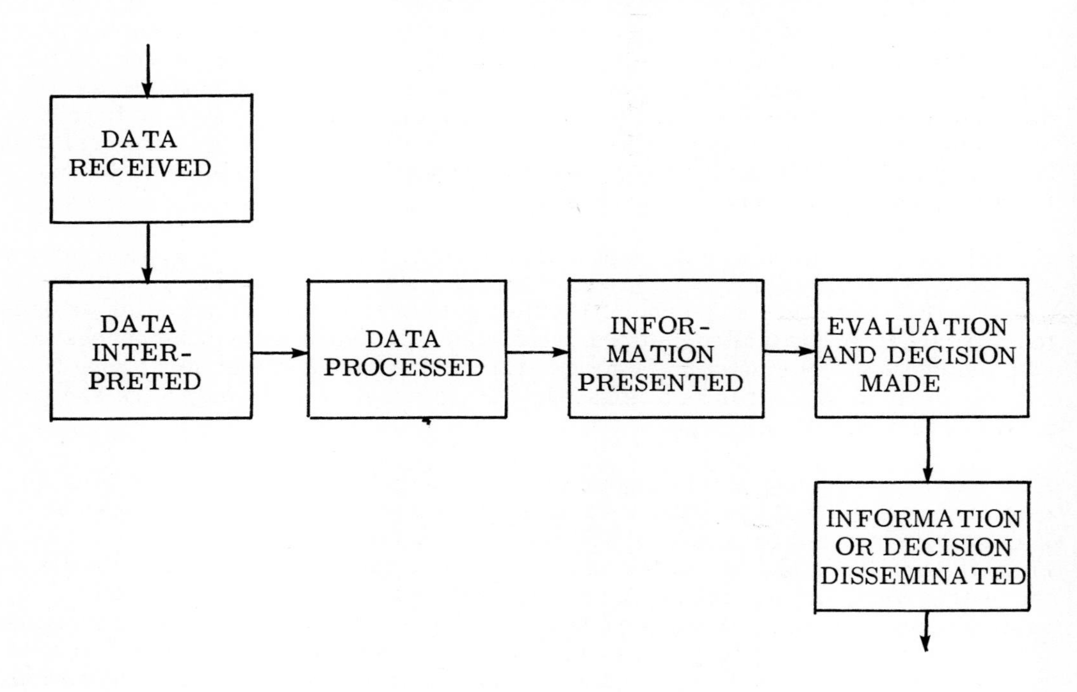

Fig. 7.3 Data, information and decision making

It has also been found extremely difficult to decide whether the aim is to provide a single fully automated, integrated system or whether it is preferable to develop independent operational modules in stages. The latter option has, of course, a lower initial cost and experience in the use of each system component as it is introduced into service can be incorporated into subsequent development of other components. A pre-requisite of proceeding in this way is that interfaces between system components must be defined in detail at a very early stage in order that interoperability can be assured for the future. With the same aim it is sensible that a single agency should be made responsible for the management of the interface between the various components both during development and deployment.

Careful study of many requirements ranging from logistic applications to weapon system command and control reveals that, whilst each requires certain unique facilities, many key features are common to all. They include:

A capability to collect data from a variety of sources; for example, surveillance elements and forward units.

Facilities to aid correlation of this information and to allow the compilation of data files appropriate to the application.

Simple access procedures to allow users to display quickly concise information relevant to the task in hand.

Facilities to disseminate quickly orders and briefing information.

Resilience to failure of communications paths and system elements.

Enhancement capacity.

The database would probably include various categories of data including first, global data which is distributed to all processors and available to all users. Then group data restricted to a defined group of processors, and finally private data which must be restricted to users connected to a single processor.

Where enhancements and additional computing power are required, satellite processors may be connected to the network to provide special features: for example, powerful data storage and retrieval facilities for intelligence applications, 'on call' computation facilities for engineering purposes or fire control calculations, and facilities for ammunition and stores stock control and reordering.

Several countries are developing C^2 systems, notably FRG (HEROS), USA (SIGMA) and UK (WAVELL). All have been hindered to a greater or lesser extent by the difficulty encountered in specifying the detail of the requirement. Project WAVELL surmounted this problem by deploying commercial, off-the-shelf, equipment to an operational division to provide the staff with ADP assistance as soon as possible and to allow them to formulate the real requirement.

WAVELL

In 1966 the British Army drafted a requirement for a project called 'WINDSOR', the aim of which was amazingly simple; it was to automate command and control on the battlefield. With hindsight it can be seen that it was either a piece of visionary inspiration or else naivety. It was, unfortunately, the latter for it was not made clear at the time why there was a need to automate command and control on the battlefield.

In particular it was not clearly specified what benefits and improvements were demanded from ADP; the term 'force multiplier' has since been used to cover this aspect. Response times were not specified; it can be readily appreciated that problems would occur if widely varying response times were demanded from a common database. A firm direction on security was not given: this could result in the mixing of secure and insecure data in memory and over data communications links; at this time data integrity and ownership were very new concepts. No guidance was given on what procedural and organisational changes could be accepted. There was no consideration given to whether centralised or distributed databases should be used; as a result the problems of varying battlefield organisations and procedures for changes of command were not addressed. Finally the problems of interoperability were not addressed; it was not foreseen that interoperability could be necessary at many levels with a consequent cost penalty.

It was hardly surprising that Project WINDSOR did not progress far. Two years later, after a great deal of thought, a better outline requirement to provide a mobile command information system for British Army tactical application was produced. By then it was appreciated that ADP techniques could do much to reduce the time and communications resources devoted by the general staff to dissemination and acquisition of routine information necessary for the control of troops in the field. A family of three interlinked command and control systems was envisaged: they were WAVELL covering the activities of 1(BR) Corps in forward areas, linked to HAIG covering BAOR and the Corps rear area logistic support functions, and a UK based, RAF manned system called TRENCHARD co-ordinating the movement and resupply by land, sea, and air of British forces deployed overseas. In the event this, too, was an over ambitious plan and only WAVELL survived. Its aim was to give selected staff cells in 1(BR) Corps and its Divisions early limited assistance in C^2.

It was to use current technology and relevant research to provide continuous experience and was to evolve, with user assistance, a system which satisfied the requirements within specified financial limits. An early feasibility study concluded that the most sensible approach lay in extensive practical trials; this is referred to as a 'prototype approach'. A pilot system was developed to automate selected staff procedures with a view to gaining practical experience in the field; it consisted of an evolutionary development programme to establish the user's requirement, or 'suck it and see'.

This scheme was planned to be in two phases. Stage 1 was to be introduced into a single armoured Division for trials in 1978 and it included the complete element for Corps HQ. If this proved successful, implementation of Stage 2, which was an increase of the quantities scaled to cover the whole Corps, would follow. In the event Stage 1 exceeded all expectations in conceptual terms but the hardware was insufficiently robust because it consisted deliberately of commercially designed hardware. It was therefore decided to provide special-to-purpose militarised hardware and to delay the issue until the early 1980s.

Two main types of WAVELL installation were provided in Stage 1. Corps and Divisional HQs were provided with a 4 tonne air conditioned vehicle containing a militarised PDP 11/34 with core store, magnetic disc and a magnetic tape cartridge unit, together with VDUs, hard copy printer, and all the necessary interface units. A schematic of this is shown in Fig. 7.4. Brigade HQs had similar facilities mounted in a long wheelbase Landrover but with fewer access facilities. Most of the hardware could be dismounted up to 400 m into tents, buildings, or other vehicles.

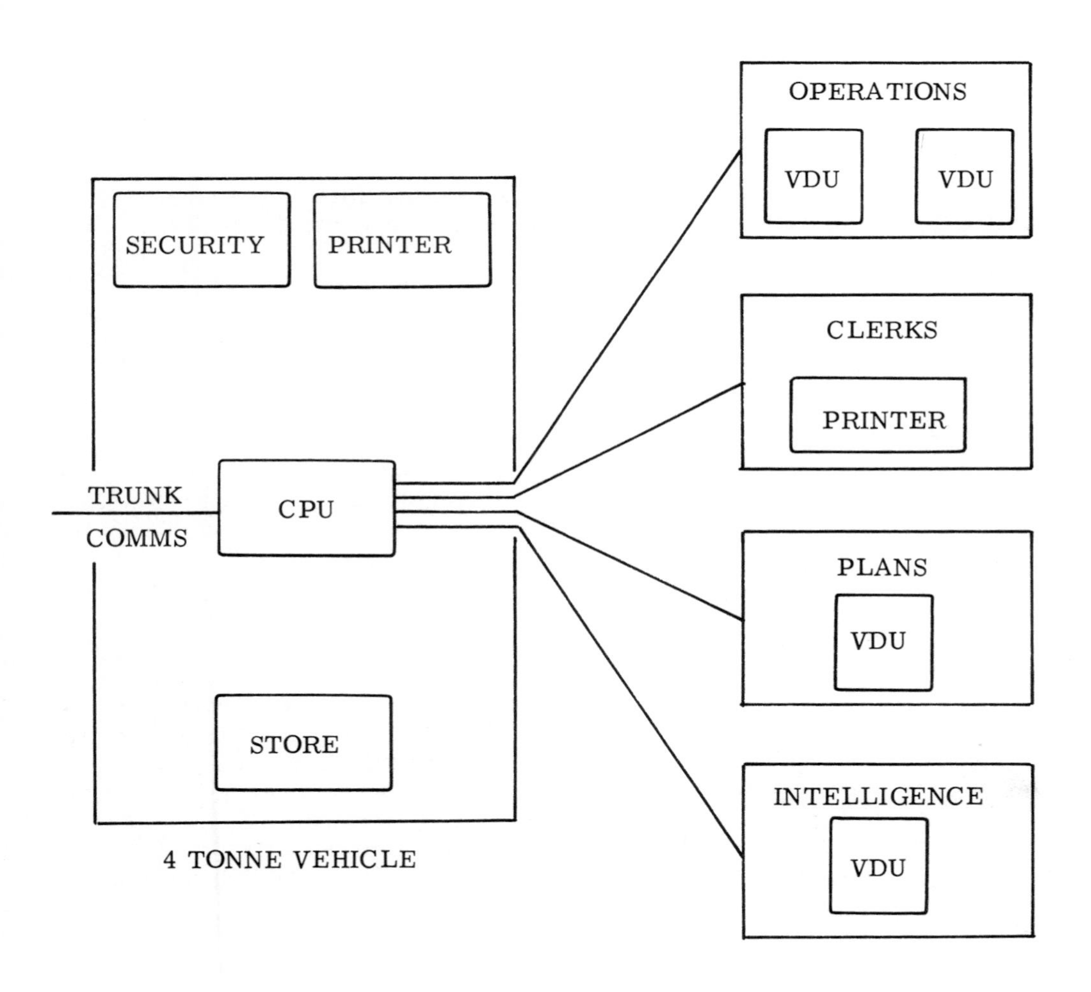

Fig. 7.4 Schematic of a WAVELL installation

WAVELL's configuration is that of an information storage and retrieval system with a distributed database. It automatically stores and distributes data inputs via trunk communications from individual subscribers to all others on the trunk system. The manual information recording processes are reduced to a minimum and the use of data transmission relieves the loading on the voice channels of command radio nets.

Data can be rapidly retrieved from the database. Consequently a change of command between two headquarters requires only that the new headquarters with command is connected in to the WAVELL network for rapid update of its database to be possible. This facility results in far less disruption to the command function than in a manual system.

System and applications software written in CORAL has been developed in parallel, so reducing integration time. The database is capable of expansion to facilitate software and hardware enhancements.

Fig. 7.5 WAVELL computer rack

The principal improvements for Stage 2 are that some items of hardware have been replaced. The CPU is now a GEC 4000 series processor; the VDU is a plasma panel development; the backing store is provided by bubble memory. The complete hardware installation has been designed around robust airportable racking. Two floppy discs are included, one to carry baseline data and the other for archiving. Figure 7.5 shows the WAVELL computer rack and Fig. 7.6 shows how an individual box can be pulled out for maintenance.

Fig. 7.6 WAVELL tray withdrawn for maintenance

Figure 7.7 shows the interior of a specially designed 4 tonne vehicle with the ADP rack on the left; the printer and keyboard are on the right. There is also a requirement to provide a vehicle with a data channel switch, consisting of a processor and a program loading unit, at each communication centre to switch WAVELL data via a multiplexer into the trunk communications system.

Selected staff cells at Corps, Division and Brigade HQs are provided with VDUs and hardcopy printers. Each HQ has its own database comprising data input by its own staff and other data provided by other cells. Much of the data in the database is 'system generated'; for example items of intelligence are compiled into groups of information. The basic concept for the storage, retrieval and display of information is by the use of a number of pre-defined FORMATS, or page-displays, of information. For example there is a format for unit data with a page-display of all the information the staff are likely to want to know about any battle group under command - such as locations, attachments and detachments, AFV strengths and combat effectiveness. The display of a format is achieved by the keying-in of a two digit number while hard copy printouts are obtained by pressing a button marked, simply, PRINT.

Fig. 7.7 WAVELL vehicle interior

Figures 7.8, 7.9 and 7.10 show the available base formats, an example of a UNIT LOCATIONS format and a chart of the intelligence formats.

BASE FORMAT

OPS.. BASE FORMAT ALT IDS INT.. PLANS GP... SD... ....

001	W	EQPT INDEX	100	W	EN EQPT INDEX	200	W	GP LIST
002	W	FMN INDEX	101		EN FMN/UNIT INDEX	201		GP LIST INDEX
003	W	HQ DATA	102	W	EN START STATE	205	W	MESSAGE
004		HQ SUM	105		EN ORBAT	206		MESSAGE INDEX
010		UNIT INDEX	106	W	EN REGROUPING	210	W	LOG UNIT LOCSTAT
011	W	UNIT DATA	110	W	EN HQ DATA			
013		BG LOCSTAT	111	W	EN UNIT DATA			
014	W	SP UNIT DATA	113	W	EN AFV/GUN STATE	250	W	REAL ESTATE CODES
016	W	REGROUPING	115	W	EN ?FMN/UNIT DATA	251	W	REAL ESTATE DETAIL
020	W	COMBAT STRENGTH	116		EN ?FMN/UNIT SUM	252		REAL ESTATE SUM
021		SP COMBAT STRENGTH	120	W	EN CONFD SOURCES			
022	W	FMN DATA	121	W	EN CONTACT REPORT			
025		ORBAT	122		EN CONTACT SUM			
030	W	STRIKE WARN	125	W	FLET			
031		STRIKE WARN SUM	126		FLET LIST			
032	W	RES DML	130		EN OPPOSING			
033		RES DML SUM	131		EN OPPOSING LIST			
035	W	SITREP	140	W	INTREP			
036		SITREP INDEX	141		INTREP INDEX			
037	W	ORDERS	142	W	INTSUM			
038		ORDERS INDEX	143		INTSUM INDEX			
041	W	CODEWORD INDEX						
046	W	NICKNAME INDEX						

Fig. 7.8 WAVELL base formats list

WAVELL has a good deal of resilience due to the fact that each database at each cell is almost identical or 'replicated', although it has been appreciated that not all cells need to hold all the information. This replication is available because the WAVELL system allows most messages to be sent automatically to a database without operator intervention at the receiving cell, so avoiding unnecessary queues. Other nations' systems under development set up individual message transmissions and each cell has a different database depending on whether the receiving cell decides to accept the data offered or not. It is difficult to know, in the latter case, whether the reason for the marked operator intervention is by design or due to the limitations of the communications system available.

WAVELL is designed to operate over trunk communications. Providing a mobile extension to the trunk system can be produced, albeit a single channel only which would impose time penalties, there is no reason why there should not be access to a processor in areas forward of Brigade HQs for data entry devices. Some form of intelligence or processing power could even be added. Similarly, should technical progress be made on such items as interactive tactical map displays these could easily be added to the system which has sufficient 'hooks' on which to hang them. In its initial deployment WAVELL provided ADP assistance to operations and intelligence cells only. Other functions, such as logistics, could be easily included should the requirement prove sufficiently important to justify the cost.

The benefits of WAVELL have been proved. Immediate access to a common, automatically updated, database by any cell in the network including step-up HQs has been a great advantage. Data transmissions have provided more accurate information and better success over poor circuits. The sending of reports and returns automatically has eliminated much of the watch and log keeping task.

C^2 systems are expensive but it is easy to understand why they are so important. Most nations can afford to maintain only a small army, comparatively speaking, and every effort must therefore be made to make it as effective as possible. Good command and control can reduce response times and labour intensive tasks and possibly save manpower, thus providing more men to fight. There are, of course, various ways of implementing C^2 projects to achieve these aims. WAVELL reflects the British concept of maximum autonomy of control at all levels. It provides the ability to be far more flexible in the delegation of responsibility, as the situation evolves, than was possible with a manual system.

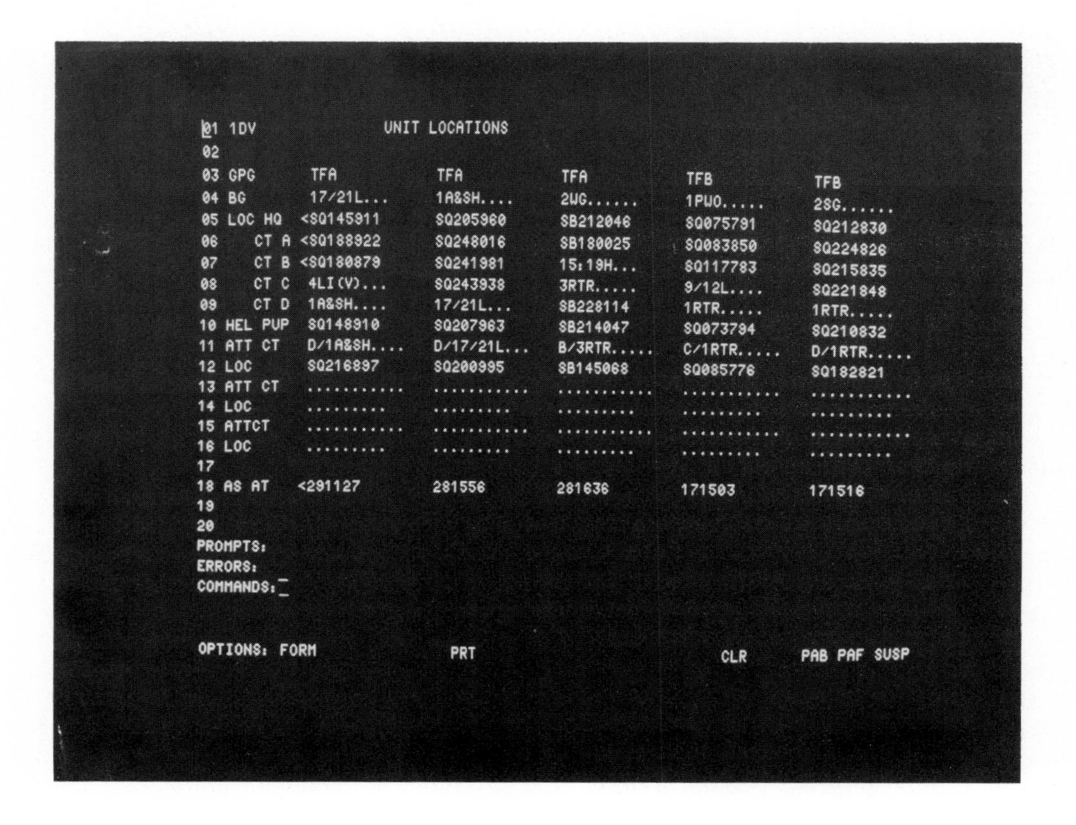

Fig. 7.9 WAVELL unit locations format

By making common up-to-date information available to all users on demand, the staff are able to spend a much greater proportion of their time on the evaluation and analysis of information, which is the basis of decision making. WAVELL is, however, a first generation system and does not integrate automatically any surveillance or acquisition devices. Nor does it make decisions. These enhancements, should they be necessary, will evolve with user experience.

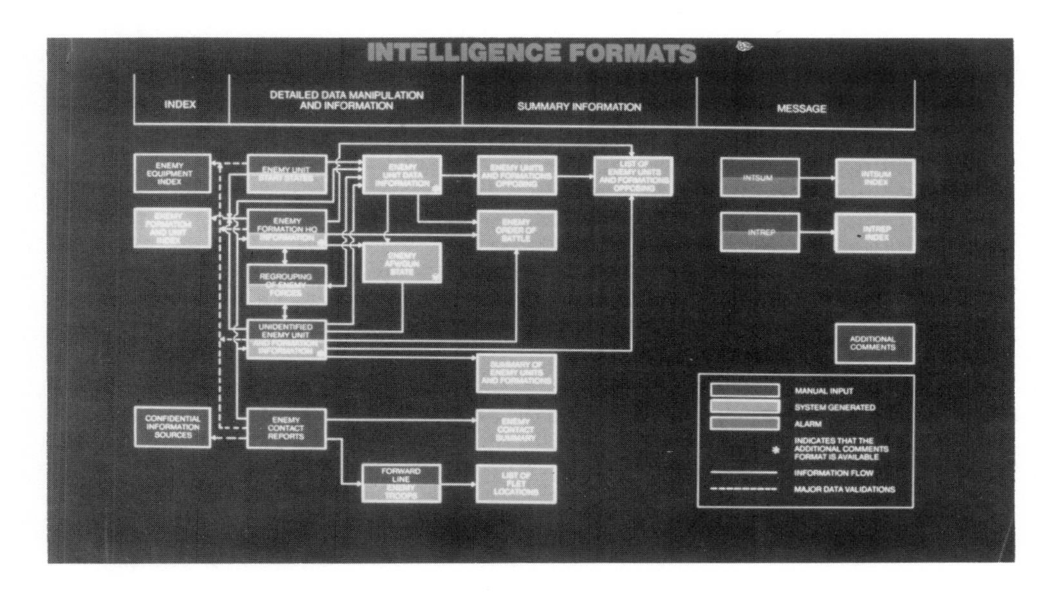

Fig. 7.10 WAVELL intelligence format chart

THE SURVIVABILITY OF COMMUNICATIONS

All new weapons, including their electronic packages, must be carefully examined to ensure that they can survive on the battlefield. ADP and its associated communications are of particular concern because of their vital role in command and control. It is the communications system which conveys the commander's orders that result from his decision-making process, and just as important, carry information back to his headquarters from individuals and surveillance sensors. The latter control information allows the commander and his staff to react to the enemy and improve the original orders. Efficient, ADP based, command and control systems are considered to be force multipliers: they justify the expense of complex weapons by enabling them to be employed to greater effect than in the past.

Technical Problems

The threat must be the criterion when assessing the survivability of a battlefield system. The potential enemy in any future war has policies, procedures, and equipment designed to minimise the strength of an opponent. Radio Electronic

Combat (REC) is just another part of a commander's tactical plan and the enemy will use it. This is reflected in normal defence planning in which armies must expect to operate their command and control under conditions of active enemy electronic countermeasures. A number of other technical pressures or threats to battlefield C^3 systems exist. A particular problem is the overcrowding of the radio spectrum. Then limited manpower and training time demands equipment which is simple to operate, reliable and easily maintained. In such alliances as NATO it is necessary to communicate with allies. High volumes of information from new sensors and distributed automatic data devices will put more pressure than ever on the system. The high mobility of future operations will make even greater demands and the severe battlefield conditions necessitate rugged and nuclear hardened equipment. Finally the high rate of technological advance and the cost of both equipments and spares force compromises in capability, modular design and interchangeability of components.

VHF Combat Net Radio

Combat net radio will be the only feasible communication system in mobile operations at unit level. At formation level trunk communications are possible and net radio could become an alternate rather than primary means of communication. Most modern families of radios are of modular design adding supplementary modules to the basic transmitter/receiver module to produce manpack sets or vehicle sets. Common design minimises the problems of operator training, spares support and maintenance. The sets are simple to operate, rugged, reliable, efficient and both lighter and smaller than the sets they replace. In future computers and their associated VDUs, printers and power packs will no doubt be produced in a similar fashion.

Although on-line encryption for secure speech is relatively easy at VHF, because of the wide bandwidth, most radio sets will lack this facility. Consequently the enemy will be able to obtain more information. Codes will often be available for manual encryption and can actually abbreviate transmission time although more message preparation time is involved and thus makes their use impractical in periods of high activity. Data, on the other hand, is far easier to encrypt and both software and hardware encryption devices are becoming available.

Trunk Communications

Modern trunk communication systems are secure, digital, and computer switched. Yet these and other characteristics do not necessarily guarantee survivability. It is difficult, if not impossible, to eliminate all errors in the software before any computerised system is brought into service; the management software of a trunk communication system can be designed to check continuously for faults and re-route message traffic automatically but experienced operators will still be required to deal with unpredictable faults. This critical area highlights the need for meticulous management of software development and testing.

The hardware should, of course, be quite reliable with proven design, BITE and nuclear hardening. A switching device should, for example, be easy to operate.

It could display error routes for any path established, then provide visual identification of any part of the system at fault, allowing a new assembly to be plugged in for a quick and simple repair. The switch must also be powerful enough to handle high volume traffic without failure.

ARTILLERY ADP SYSTEMS

For some years most of the world's advanced artillery systems have used ADP assistance to compute ballistic, meteorological, and survey problems. Indeed the requirement in the USA for a computer to produce firing tables during the 1939-45 war was the main reason for the development of the ENIAC which was a milestone in the history of computers. Some examples of applications are described below.

FACE

The UK Field Artillery Computing Equipment (FACE) was brought into service in the 1960s. It is the most widely sold mobile artillery computing system and is used in artillery command posts and survey computing centres to automate the preparation of gun firing and survey data. FACE is a good example of the successful deployment of fairly sophisticated processing capabilities on to the battlefield. The methods of fast, automatic computation employed require less skill to operate than previous manual systems. The simplicity of operation also considerably reduces the risk of human error and training times.

The computer system was originally based on the Elliot 920B but is now available with the more modern MC1800 microprocessor which is faster, more reliable, smaller, and has a far greater memory capacity. Together with its control console, it can be installed in most wheeled and tracked vehicles commonly used as command posts. It is built to operate within the normal range of military specified climatic conditions.

In the gun data role, all the information necessary to ensure that a shell or rocket arrives quickly and accurately on target is processed by the computer and displayed on the operator's console. For each calibre of weapon there is a program tape stored in a sealed cassette: the data can be fed into the computer in about two minutes. FACE can operate with all known artillery equipments and programs are constantly being refined and extended for new types of weapons. Changes and variations in gunnery procedures can also be dealt with by amending the programs.

Lttle routine maintenance is necessary and a test sequence in each program provides the operator with a confidence check. Complete test programs and test sets exist for diagnostic testing of all components. They isolate faults to particular sub-units.

Fig. 7.11 FACE display

For such an early operational field computer system certain man/machine interface aspects of FACE are very advanced: the console consists of a matrix of coloured windows, in an illuminated panel, and a functional keyboard. Each window relates to a particular stage in a computation and illuminates in sequence to lead the operator through his inputs to the keyboard. (See Fig. 7.11).

AWDATS

The Artillery Weapon Data Transmission System (AWDATS) is an equipment developed in parallel with FACE to display the firing data, computed by FACE, at the individual guns of the firing unit. The data for each gun will vary due to the different location of each gun within the battery and variations of muzzle velocity.

The guns are fitted with a data display unit which receives data from the command post. The display unit can be fitted to the frame of a towed gun or installed in the vehicle for self-propelled guns. Transmissions of data are by FM audio signals over net radio or line.

Fig. 7.12 AWDATS display

The advantage of AWDATS is that, by partially eliminating the requirement for verbal transmissions and the checking back of gun firing data, the speed of reaction and thus the efficiency of the artillery system is greatly improved. (See Fig. 7.12).

AMETS

The Artillery Meteorological System (AMETS) is a self contained mobile meteorological station built to military standards. The system is designed to provide rapid, up-to-date meteorological information for artillery firing units in the field. The requirement for such systems has been highlighted by the introduction into service of equipments such as FACE. In order to achieve the first round accuracy, which is now possible, the artillery system requires fresh and accurate 'met' data.

The data processing part of the system uses the same proven 920B central processor and power supply equipment used in FACE and the man/machine interface is a development of the successful console interface used in FACE. The WF3M wind-finding radar is an auto-tracking unit developed from the widely used civil WF3 wind-finding radar. The radio sonde, however, is a special military design and requires no pre-flight calibration.

AMETS consists of an instrumentation vehicle, a command post vehicle, radar trailer and ancillary stores vehicles and trailers (see Fig. 7.13). The system can produce on demand any combination of the following meteorological messages:

Computer message on punched tape which can be fed directly into a FACE system

Standard ballistic message for use by units not equipped with FACE

Drone surveillance message

Sound ranging message

Nuclear fallout message

Civil forecasting message containing raw met data

Fig. 7.13 AMETS vehicles and radar trailer

AMETS is easily and rapidly deployed making it suitable for use in the same areas as artillery batteries. The use of automatic data gathering techniques and processes eliminates the long delays and chance of human error associated with manual meteorological computations. Once the balloon-borne radar reflector has been acquired by the radar the system runs entirely automatically.

MORCOS

MORCOS, shown in Fig. 7.14, is a lightweight hand held computer for use at mortar positions to produce fast, accurate firing data for mortars. It is simple to operate and reduces the chance of error in what is normally a very unfavourable operational environment. Improvements in modern mortars, their ammunition and the means of acquiring mortar targets have meant that the remaining delays, errors and weaknesses in the system feature more strongly in the method of producing firing data. ADP equipment can do much to ease the problem by replacing the time consuming and often inaccurate manual methods of fire prediction.

Fig. 7.14 MORCOS

MORCOS consists of a single self-contained unit incorporating a processor, a keyboard for data entry, and display and batteries as shown in Fig. 7.15. Its programs provide data for high-explosive, smoke and illuminating missions. It also deals with survey problems and target reduction. Finally, it can cater for up to three shoots in parallel. It accepts inputs to take account of the variables which affect the accuracy of mortar fire such as meteorological conditions, charge temperature and differences in altitude between mortar and target. Observers' corrections can be easily applied and it will store information on up to ten mortar positions, fifty targets, ten forward observer locations and nine 'own troops' positions.

Fig. 7.15 MORCOS display

The processing is carried out by a microprocessor with associated semi-conductor backing store. Several ballistic programs exist and each can predict for one type of mortar and its complete range of ammunition. Programs can be quickly changed by removing a plug-in module and replacing it with another.

The keyboard is waterproof and there is an electroluminescent panel under the keyboard which can be illuminated by a pushbutton for night operation. It has only twenty-four keys, ten of which are for digits. The display has eight alpha-numeric characters and uses light-emitting diodes (LED) with a variable brightness control.

MORCOS can be powered by batteries or by an external power source. It weighs, including batteries, less than 1.35 kg and measures 230 x 110 x 54 mm. The minimal use of internal electrical connections, due to the employment of LSI techniques, makes the system very reliable but should a fault occur sub-assemblies can be easily replaced and no special first line test equipment is necessary.

BATES

There have recently been many improvements in target acquisition devices and improved artillery weapon performance. In order to optimise the new capabilities, and to overcome any imbalance in relative firepower, it is necessary to co-ordinate the acquisition devices, weapons and computing aids into an efficient command and control environment. Such a system is being developed in UK in BATES: it is intended for service later in this decade.

BATES stands for 'Battlefield Artillery Target Engagement System' and is an artillery command and control system for the co-ordination and control of in-direct fire. It consists of ADP equipment and programs to enable artillery commanders and staff to make the most effective use of their artillery resources. BATES is a Corps level system with distributed processors at Battery, Regimental, Divisional and Corps level command posts.

Passage of information is achieved by digital transfer of data over combat net radio, line, and trunk communications. Component equipments consist of processors, VDUs with keyboards, graphics display units, map overlay plotters, printers, and data entry and readout devices.

The requirement for BATES stems from the need for an efficient system for the allocation of limited artillery resources in the face of an increasing threat. Such an overall system must embrace target acquisition, collation, storage and retrieval of data, data analysis, data transmission, fire planning, ballistic computation and logistic accounting. BATES will help by providing automation of the following field artillery functions:

Resource control, deployment and allocation.
Fire control.
Firing unit and ammunition status reporting.
Fire planning.
Artillery target intelligence gathering, collation and analysis.
Nuclear target analysis.
Target records updating.
Meteorological data distribution.
Reports and returns handling.

In this way it will improve artillery performance in several areas. There is a clear requirement to engage the highest priority target with the optimum weight of fire at the right time; the greatest effect can be achieved by burst fire or the concentration of the fire of several batteries. The artillery commander will be able to effectively allocate his various resources to suit the rapidly changing battle situation; this flexibility is not available in a manual system. Ballistic applications software will allow mixtures of calibre with no restrictions on the dispersion of the guns.

Communications will be much faster and will be in digital form but some voice nets may be retained to give a back-up capability. Facilities for four communications modes are provided; they are time division multiplex (TDM), normal contention, voice led data and voice only. TDM allocates a time slot to each station for his transmissions, guaranteeing access to the net: the cyclic rate of the time slots, however, means that complete messages are sent and received in near real-time.

The introduction of BATES does imply a loss of most of the traditional voice nets and a corresponding loss of the inherent "all-informed" characteristic achieved by staffs listening to messages being passed between other stations on the net. The advantages of digital transmission lie in the reduction of net occupancy and elimination of error-inducing manual intervention; accuracy will be much improved by the employment of E D and C methods and consequent automatic acknowledgement of messages which are correctly received. Voice nets simply would not cope with the projected volumes of message traffic.

The large number of cells deployed at each level demands as much commonility of hardware as possible. Common hardware throughout the system will also allow cells to call upon the hardware of other cells in the event of equipment failure or loss. Thus a small range of very versatile equipment is being produced. The hardware components are:

CPUs - microprocessors
Semiconductor store
Plasma panel VDUs - some with graphics capability
Program loading and data recording units
Printers
Graph plotters
Data entry devices (DEDs)
Net radio interfaces
Line interfaces
Trunk interfaces
Gun display units (GDUs)

Forward Observers and certain other cells will be equipped with a Data Entry Device (DED); this will be a portable entry and output device which normally will use fixed formats; the observer will complete the format, which will then be transmitted automatically. The device will contain a communications microprocessor, for message handling and for automatic acknowledgement of messages; it will interface with net radio, trunk or line.

Guns will have a compact Gun Display Unit (GDU) which will receive, acknowledge and display all firing data.

The battery command post will have an applications microprocessor to replace FACE and this will cater for mixed batteries whether by calibre or by ammunition type and will allow for a widely dispersed gun position where necessary. The operator will use a VDU and alphanumeric keyboard and also a printer for making hard copy of messages. A space model is shown in Fig. 7.16. The battery commander also will be equipped with a processor, VDU and printer to enable him to form a Fire Support Co-ordination Centre (FSCC) when necessary for the overall co-ordination of the forward observers and the guns.

Fig. 7.16 BATES space models in an AFV432

Key:		
	VDU	2 Visual Display Units
	Keyboard	2 keyboards
	SCRA	Single Channel Radio Access (Trunk)
	NIE	Net Interface Equipment (CNR)

Major command and control cells such as the Fire Direction Centre (FDC), Tactical Headquarters (Tac HQ) and Artillery Operations (Arty Ops) cells will have processors, VDUs for current work and for planning, a printer and a map overlay plotter. A large number of messages will go directly into memory and need not be seen by an operator (see Fig. 7.13). Artillery intelligence cells will share the Arty Ops processors but will have their own VDUs. Artillery commanders will have a printer and VDU in their rover vehicles. Meteorological command posts will have BATES equipment to replace the ageing AMETS hardware.

BATES is part of the developing UK Army C^3 network. The PTARMIGAN trunk communications system will provide a loop system within each headquarters and nodes across the battlefield. At each HQ there are terminals from several ADP systems interfacing with the loop and thus communicating with each other; in this way BATES and WAVELL can exchange information. PTARMIGAN will also provide Single Channel Radio Access (SCRA) facilities to selected subscribers (for example the CRA in his mobile role).

Superimposed on this trunk system is the range of CLANSMAN combat net radio (CNR) sets provided for the artillery. CNR is the primary means of communication below Brigade level and provides parallel communications with PTARMIGAN at Brigade and above. BATES will be deployed at all levels and will interface with PTARMIGAN via the loop at each formation HQ, using SCRA where applicable. All BATES cells also communicate using CNR and in most cases this provides a vast potential for flexibility.

During a feasibility study the questions of centralised or distributed processing and databases, local or remote storage, deployment of hardware with users or at communication nodes, and degrees of redundancy and duplication were all addressed. Influenced to a large degree by the constraints of the British Army artillery organisation and the type of communications planned to be available in the BATES timeframe, the recommended solution was to distribute processing to the user cells of the current hierarchy, provide local storage, and to rely mainly on the current system of back up which utilises pre-selected step-up and alternative HQs.

BATES employs a common microprocessor which will provide the computing power and facilities required in the larger cells but is of sufficiently low cost and electrical power requirement to be feasible at the bottom end of the range as well. The advent of high density low power semi-conductor memory also allows a common single level modular storage medium throughout and each computer assembly will allow some 1.5 megabytes of storage. Should the higher level cells require a larger database then the system can be extended to an additional assembly holding storage cards only.

The VDU is highly intelligent and capable of dealing with all local formatting and editing tasks associated with message composition. Its processing and storage will be provided by standard BATES components and the display head is based on a 512 x 512 point plasma panel, a keyboard, and an interface card dealing with the keyboard, processor, and manual control interfaces; some VDUs will have a graphics capability. The teletype will be the one described in Chapter 2 and issued as a standard item of equipment in the British Army. The program loading unit will have an extra function; it will also act as a data recorder and is

for the technological improvements which have led to the appearance of VLSI electronics such as the microprocessor.

Artillery Computations

Three artillery data systems which make use of microprocessors are FACE, BCS and BATES. FACE and BCS use bit-slice microprocessor circuits to build the fast central processing units which form the nucleus of each installation. In the case of FACE the early versions used a minicomputer with an 18 bit wordlength. By using five 4 bit-slices it was possible to build a bit-slice microcomputer which emulated the original minicomputer, although very much smaller physically, and so all the existing software could be used without modification. In FACE and BCS it would be impractical to use a single-chip microprocessor as the basis of the microcomputer because its data processing speed would be too slow. One way in which single-chip microprocessors could be used, though, would be to have a number of such processors within each computing unit and share tasks between them. In this way their effective speed would be improved.

BATES is based on a series of common modules which make use of 16 bit single-chip microprocessors. Where high speed is required, such as in ballistic computations, a number of microprocessors are used together. As mentioned earlier the BATES common module types include CPU, RAM, ROM, display drive, BATES input/output, and CNR and trunk interface. The main computers used in BATES are built with three CPU modules to provide the required processing speed. Other units such as digital entry devices and gun data displays make use of single CPU modules, since they do not have to perform so much data processing. The CPU modules also incorporate BITE; this consists of software routines which are designed to exercise and test the processor in order to detect failures or errors as soon as they occur. It minimises the chance of complete system failure or the transmission of incorrect data. The BITE facilities can then aid maintenance by the identification of faulty items: immediate repair by replacement can be carried out. Also the multi-processor nature of BATES means that if a module fails, its tasks may often be taken over by another. Consequently, although speed may suffer the system does not fail: this is termed 'graceful degradation'. It is likely that the trend towards handling ever larger amounts of data will result in a growing number of multi-processor systems appearing in military equipment, in contrast to the centralised single main processor approach used in earlier systems such as FACE.

Tank Fire Control

The fire control systems (FCS) of most modern main battle tanks make use of microcomputers to perform ballistic calculations and thus determine the super-elevation and lead angles required to engage a target. These systems typically operate in the following way. The gunner identifies the target and selects the type of ammunition to be used. He lays his sight onto the target and presses a button to operate a laser rangefinder whilst continuing to track the target if it is moving. The laser rangefinder provides a range output to the processor, and rate sensors provide information on target motion; this is obtained whilst the target is being tracked. A tilt sensor attached to the gun trunnions feeds trunnion cross tilt data to the processor, and additional inputs are available for charge temperature and wind data. Further possible inputs include features such as barrel wear. The processor reads all the inputs and then performs a ballistic calculation to find the super-elevation required for that range and ammunition, and then from the computer transit time can derive the lead angles necessary to allow for target motion. The system injects a spot or other type of aiming mark, suitably off-set to take into consideration all the inputs, into the gunner's eyepiece. He lays this on to the target and fires. Systems of this type greatly improve the hit probability against both stationary and moving targets, and the time taken to engage is significantly reduced.

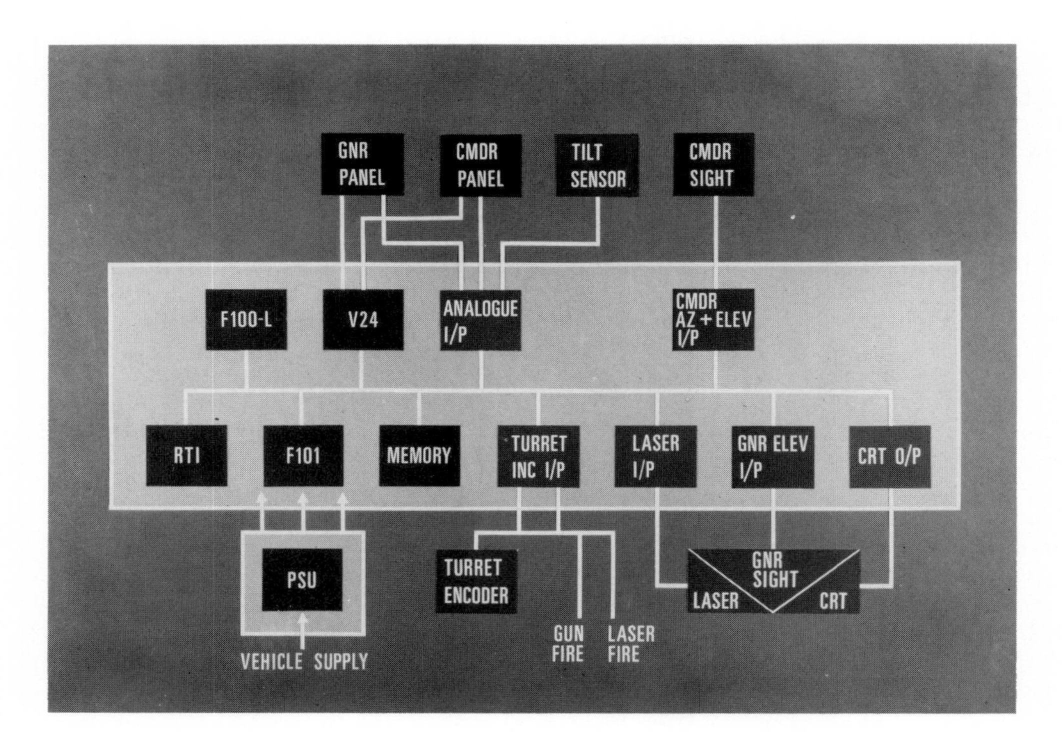

Fig. 7.20 FCS block diagram

Key:

F100-L	16 bit microprocessor
F101	Multiply/divide i.c.
V24	Serial interface
RTI	Real-time interrupt (generates signals at constant time intervals)
CRT	Cathode ray tube (generates aiming mark)
PSU	Power supply unit

The block diagram of one microcomputer-based FCS is shown in Fig. 7.20. Figure 7.21 shows the actual hardware of another such system; in this case one intended for retro-fitting to vehicles such as Centurion and M60.

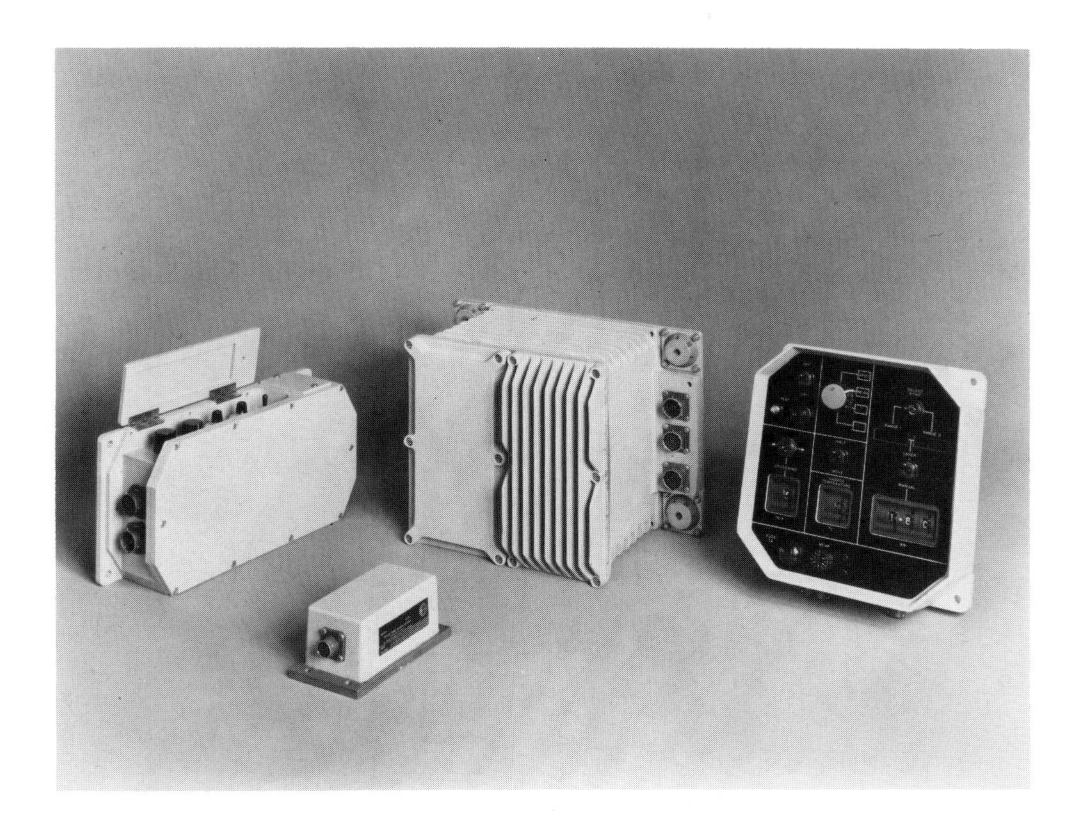

Fig. 7.21 FCS hardware (MRSL SFCS - 600)

Figure 7.22 shows the processor box from the latter system, with its cover removed and plug-in CPU board extracted. The half of the box still sealed contains the power supply, which is as large as the microcomputer section. The CPU board, based on an 8 bit microprocessor, can be seen to hold a number of EPROM integrated circuits, identifiable by the windows in their top surfaces. These contain the information on the ammunition types and gun in use. If it is desired to fit the system in a different type of tank, or change the ammunition types, then it is only necessary to change the plug-in EPROMS. Systems intended for use in environments where radiation may be present may however use mask-programmed or fusible-link ROMs, or core store instead, since the data held in EPROMS can be erased not only by ultra-violet light but also by ionising radiation.

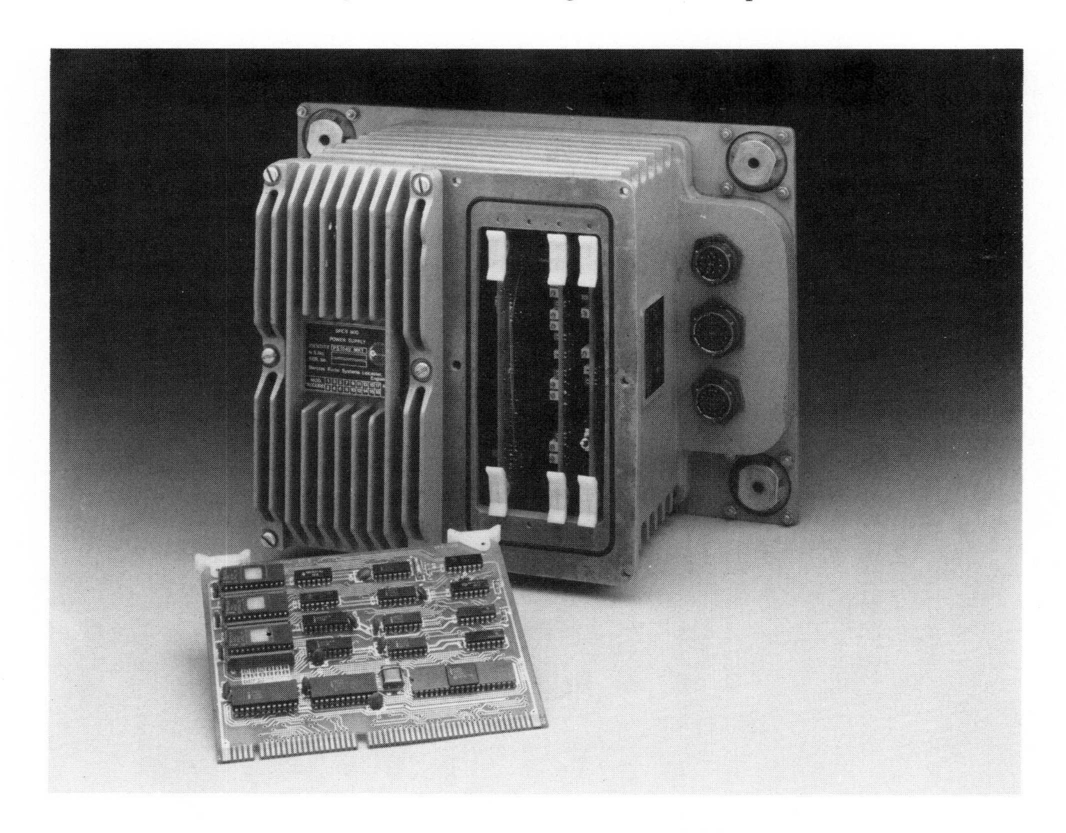

Fig. 7.22 FCS processor box (MRSL SFCS - 600)

Artillery Sights

A major, new application is the use of a microcomputer within the Indirect Fire Sight (IFS) of a self-propelled howitzer in place of the mechanical linkage and compensation previously used to overcome the effects of gun platform tilt on accuracy of lay. The mechanical systems were of course acting as mechanical analogue computers, and they have now been replaced by an electronic digital computer making use of various sensors and used in conjunction witha 'fixed' (non-levellable) sight. Two 8 bit microprocessors are used in this Electronic Plane Conversion (EPC) sight. One is used to interface the sight to whichever data transmission system is in use (eg BATES), since this varies depending upon the country using the system. The second performs the co-ordinate transformation in order to calculate the correct lay of the tilted gun to meet the command bearing and elevation, which are referred to earth's axis co-ordinates.

The system is shown mounted in a mock-up of part of a turret in Fig. 7.23. The fixed sight, processor box and data display unit (DDU) are clearly visible. Inputs to the processor come from three shaft encoders. One measures sight head rotation and the other two gun and sight mirror elevation. There are also two

Fig. 7.23 EPC sight in turret mock-up

tilt sensors. These values, together with the command bearing and elevation enable the processor to calculate the correct lay in terms of traverse of the tilted turret and elevation of the gun in its tilted trunnions. The differences between these figures and the present gun position generate error displays: the layer merely elevates or depresses the gun and rotates the sight head to null these. He then traverses the turret to lay the sight back onto the Gun Aiming Point (GAP): the gun is then correctly laid. Since no sight levelling is involved the layer's task is eased, and, since the computation is very fast, accurate check laying can be performed even whilst burst fire rates are occurring.

BITE routines are provided in the EPC system and these carry out continuous checks of processor operation. In addition, if command data or gun tilt is outside the operating specification a warning is issued to the layer; and when new command data is received the layer's attention is drawn to it by flashing the relevant displays on the DDU. At start of day a test button can be used to cause a more detailed set of tests to be performed to enable the processor system and sensors to be checked: if a fault is found within the processor a diagnostic message is displayed on the DDU to identify the faulty section. Field repair then consists of replacement of the faulty module, which is shipped back to base workshops for repair.

Mortar Fire Control

A microcomputer-based mortar fire control system was described earlier (MORCOS), and Fig. 7.24 shows the unit from the rear with battery and ballistic modules removed.

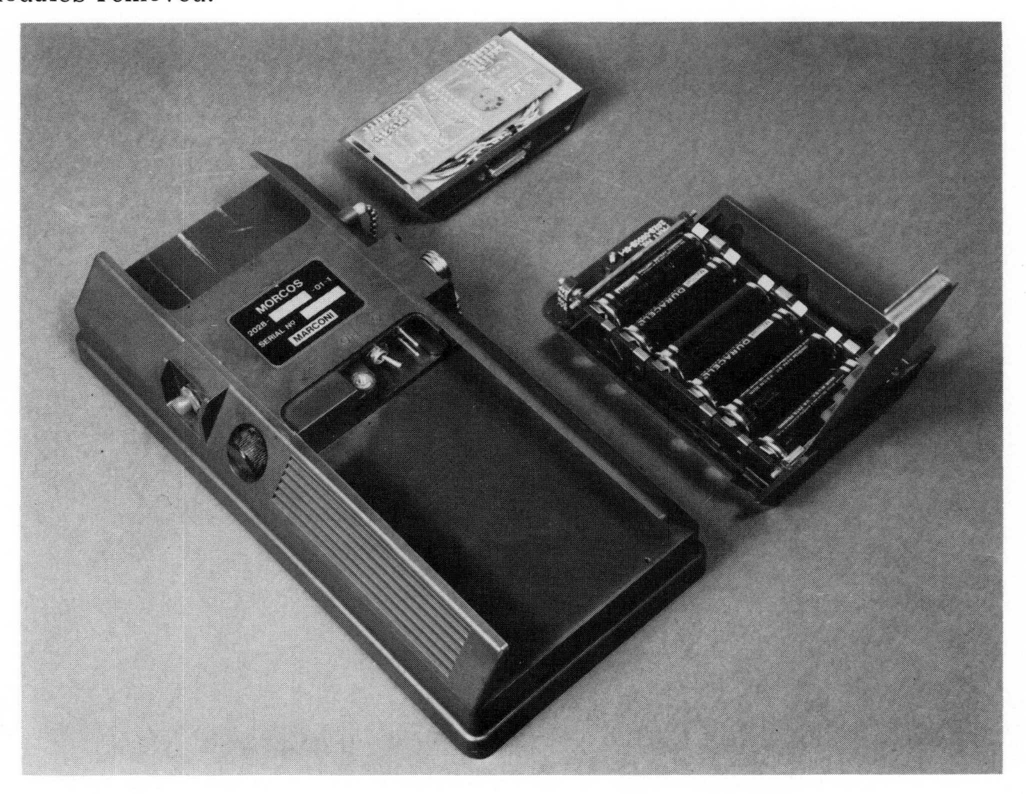

Fig. 7.24 MORCOS with ballistic module and battery removed

The interchangeable ballistic module contains information in ROM on the particular mortar type and ammunition; in order to enable the system to operate for a useful period of time from the small battery pack, a Complementary Metal Oxide Semiconductor (CMOS) microprocessor is used because of its very low power consumption.

Tactical Data Entry Devices

Microcomputers are used in many digital message devices to store and then transmit lengthy messages in bursts; they can also accept similar incoming messages. Redundant characters are inserted into the messages in order to enable the units to detect and then correct errors which may be introduced during the transmission. Many such devices also incorporate facilities for encryption so that a degree of privacy may be possible even when the messages are freely transmitted. An example of a digital message device is shown in Fig. 7.25: this unit is in service with a number of NATO countries.

SUMMARY

As new weapon systems evolve and are introduced, new uses for ADP systems and microcomputers will be found, making new generations of equipment more effective and selective. WAVELL and BATES are first generation equipments for the British army but they represent giant strides forward in command and control; it is perhaps in C^3 and logistic systems where the greatest advances will be made in the future.

Fig. 7.25 Tactical data entry device

SELF TEST QUESTIONS

QUESTION 1 In what way does the military environment influence the choice of computer components for field systems?

Answer ....

....

....

QUESTION 2 What is a real-time system?

Answer ....

....

....

....

QUESTION 3 Explain the use of BITE.

Answer ....

....

QUESTION 4 What are the main differences between the processor design approaches used for FACE and BATES?

Answer ....

....

....

QUESTION 5 What are the benefits of a "prototype approach" as used in the initial deployment of WAVELL?

Answer ....

....

QUESTION 6 What is a "distributed system"?

Answer ....

....

QUESTION 7 Why is digital transmission of data more accurate than voice?

Answer ....

....

QUESTION 8 What is "emulation"?

Answer ....

....

....

QUESTION 9 What are the benefits of multiprocessor configurations?

Answer ....

....

....

ANSWERS ON PAGE 202

8.
Future Trends

INTRODUCTION

In earlier chapters we discussed how computers have developed and how they work, and gave examples of their wide range of military application. It is important that several of the most important current and future developments should be highlighted. If one single principle is to be selected for emphasis it must be that of 'system thinking', for there is no doubt that ADP applications are still only at the threshold of their life and the majority of individual ADP systems which are procured today will be required to form part of a larger system in the future. The problems of interoperability and modularity cannot therefore be denied; the task of the management of the interface between systems can be very complex and the complexity will increase proportionally with the lack of planning of the interface. It is vital that, even if a complete system cannot be funded at one time, the overall system plan is defined and that those modules which can be afforded should be procured to fit the overall plan. There is a need, obviously, for both breadth and length of vision on the part of ADP system builders.

HARDWARE TRENDS

More Powerful Chips

The rapid increase in the possible complexity of integrated circuits has paved the way for many hardware advances. With an expectation of 10 million devices per chip by 1990 it is reasonable to expect extremely powerful microprocessors to become available, rivalling the CPUs of almost all minicomputers and even some mainframes as currently available. It is also to be expected that single-chip microcomputers will appear containing 16 or 32 bit CPUs, 32k bytes or more of memory, and comprehensive input/output facilities.

The power of the newer MPUs is already being enhanced by 'co-processors', chips which perform specialist functions, such as rapid, high precision arithmetic, in support of the MPU. The memory management circuits available for the latest MPUs also provide them with capabilities similar to those of a minicomputer for task partitioning and scheduling. 32 bit MPUs are just beginning to become commercially available at the present time, and it is likely that they will be used in many situations previously thought the domain of minicomputers. Some of the latest MPUs are designed to support many of the Ada high level language instructions and data formats more or less directly. Other 16 and 32 bit MPUs are soon to be released by the semiconductor manufacturers which will emulate the instruction sets of popular mini and mainframe computers, and some computer manufacturers are producing microprocessor versions of their standard CPUs. These new MPUs will therefore make it possible to use existing software for many applications.

Faster Chips

To reduce power consumption many MPUs are now being manufactured using CMOS technology, and this trend is likely to continue. However, when speed of processing is the most important attribute, processors based on silicon MOS transistors may not be fast enough, and research is being conducted into alternative materials and techniques. The use of gallium arsenide (GaAs) or indium phosphide (InP) as alternative wafer materials appears to offer the possibility of speed improvements of a factor of ten or more when compared with silicon. Also under investigation are 'Josephson junctions' which can be used as very fast logic elements, switching in about one thousand billionth of a second, as opposed to the 3 billionths of a second taken by the high speed logic currently used in many computer CPUs. Josephson junctions must however be cooled to very low temperatures of a few degrees Kelvin, if they are to work, since they rely on superconducting effects: it is therefore unlikely that they will be used in military systems except perhaps when no other approach can meet some vital requirement.

Optical Processors

In the longer term it is likely that much of the complex signal processing involved in areas such as radar systems may be undertaken by optical rather than electronic processors. A large amount of research into optical processors is taking place, but a number of problems remain to be overcome before they can become feasible propositions for other than very specialised applications.

Advances in Architecture

Advances are also being made in system architecture. Many systems now under development use a number of processors, and these multi-processor systems offer potential benefits in terms of fault-tolerance. If one processor fails it may be possible to assign its tasks to the other processors. A very useful advance is the development of local area networks (LANs) which are being used to link otherwise independent computers.

An interesting area of research is content addressable memory (CAM). Whilst conventional memory is accessed by specifying the address of the data involved,

CAM is accessed by specifying the data required, and the location or locations holding that data respond. This offers great potential for many forms of list processing. For instance consider the problem of identifying those army officers qualified to pilot helicopters, given a list of all army officers and their qualifications. A conventional computer system using standard memory must work through the entire list to extract the information required. A system using CAM would simply put out as an 'address' the data 'helicopter pilot' and all those locations holding data on such officers would respond, and no others. Thus the search would be very considerably shortened.

SOFTWARE TRENDS

Specialised Languages

There is an underlying trend in software in favour of aiding program design and improving maintainability. Specialised languages have been developed which are designed to allow the programmer, or system designer, to work in terms of abstract modules; these model the required system behaviour and can finally be translated to machine code for execution. Such languages differ from most conventional high level languages because they provide for much more complex operations, closer to the terms in which the requirement will be specified. They are often designed to permit the development of a complete database in which the developing program may be stored and analysed, and which can support a number of design methodologies. Languages of this type are designed to meet a variety of needs, but early implementations have tended to be used for the development of large systems rather than those which are small, embedded and real-time.

Ada

In an attempt to improve software maintainability the US Department of Defense has decided to standardise on one programming language for embedded applications. The requirements for the language were refined through a number of stages to provide features for real-time and parallel processing, and were finally issued as the 'Steelman' specification in 1978. A number of software houses submitted proposals to meet these requirements and the DOD has selected one of these, the language now named Ada, as its standard. Ada is named after Ada Countess Lovelace, Byron's daughter, who worked for a while as Charles Babbage's assistant, and who has some claim to have been the world's first programmer.

To ensure that maintainability really is improved the DOD has insisted that no compiler may be called an Ada compiler unless it meets all of the definition, issued in 1980, and operates in the defined and thus standardised Ada Programming Support Environment (APSE). The provision of APSE means that if a programmer moves from one machine and its Ada compiler to another he will find that no new operating systems or other such features need to be learned. APSE also provides a complete database for developing software. Costs may therefore be minimised in the long term. It is likely that Ada will be adopted in due course by a number of other countries as a standard language for military software.

Since maintainability of software was the major aim, Ada has been designed for intelligibility rather than ease of writing. In many ways Ada bears a strong resemblance to Pascal, Algol and the other algorithmic languages. Like Pascal, Ada is strongly typed, that is, the compiler ensures that operations are only allowed between data items of compatible types. Thus an operation involving the addition of an integer variable to one representing a character would not be allowed. This helps to ensure that incorrect programs are detected, and maintenance is eased since specialised types can be defined to suit the task being performed. For instance a tank fire control program might define a type AMMUNITION as:

```
type AMMUNITION is (HE, APDS, HESH, HEAT);
```

where the list defines the possible values of AMMUNITION. Unlike standard Pascal, Ada provides facilities for parallel processing and for synchronisation between tasks. Ada programs can call procedures (subroutines) written in other languages, and can include blocks of machine code. The language has been designed to support real-time systems, large or small, and includes addressing facilities especially suitable for small, dedicated systems such as microcomputers, as well as features which suit it to major data processing tasks.

As an example consider an Ada procedure for converting the value of a sum of money from one currency to another. For the procedure to work satisfactorily a package called IO_PACKAGE which contains routines for the input and output of data via a terminal must be available, and the program calling this procedure must include the following definitions:

```
type CURRENCY is (STERLING, FF, DM, US_$, AUS_$, YEN);

CONVERSION_TABLE: array (CURRENCY, CURRENCY) of FLOAT;
```

(Underlining a space denotes that both terms are parts of a single identifier; the absence of underlining denotes two separate identifiers. Underlining provides for longer identifiers which equate to common English). These define the type CURRENCY to consist of the enumerated set STERLING. . YEN, and a two-dimensional array CONVERSION_TABLE containing the floating point numbers which are the conversion factors from one currency to another. Thus CONVERSION_TABLE (FF, DM) will contain the exchange rate from French francs to Deutschmarks. The values in this array must have been loaded by the main program before calling the conversion procedure which follows:

```
with IO_PACKAGE;
procedure CURRENCY_CONVERSION is
use IO_PACKAGE;
SOURCE_CURRENCY, DESTINATION_CURRENCY: CURRENCY;
SOURCE_QUANTITY, DESTINATION_QUANTITY: FLOAT;
-- This procedure performs conversion of money values
-- from one currency to another. Using an input/output
```

```
-- package it reads in source currency and quantity and
-- destination currency.  It then computes and prints out
-- the equivalent quantity in the destination currency.
begin
   GET (SOURCE_CURRENCY);
   GET (SOURCE_QUANTITY);
   GET (DESTINATION_CURRENCY);
   DESTINATION_QUANTITY:= CONVERSION_TABLE(SOURCE_CURRENCY,
                          DESTINATION_CURRENCY)*SOURCE_QUANTITY;
   PUT (DESTINATION_QUANTITY);
end;
```

Fault Tolerance

The development of fault tolerant software is a strong new trend. Much current research is focussed on developing methods by which a program can recover from an error, once it has been detected. One approach is for the program to back-track to a point at which its data is known to be correct and then proceed from that point again. This type of technique can offer some protection against transient faults, but cannot overcome the effect of logical errors in the original program. Consequently research is also being undertaken into methods of rigorously validating that programs do indeed perform as expected.

MILITARY ASPECTS

Introduction

Computers are tools of great power p-aced in the hands of soldiers who need them for a vast array of command and control and weapon control applications. Whether today's soldier can be trained to operate and maintain these tools is a fairly commonly expressed concern. The same kinds of concern have been expressed throughout history for the modern weapon or procedure of the time, from catapult to cannon and from horse drawn artillery to SP guns. Most countries engaged in the procurement of computers for defence purposes are well aware of the problem and are very actively engaged in the analysis of the training problems.

Education and Confidence

In very few fields does education 'happen overnight', neither can education and confidence in equipment progress without the involvement of the trainee. It is difficult to educate or train an individual before sufficient equipment is available for that purpose. Thus to talk of problems that might be encountered is very subjective; this does not mean, however, that a great deal cannot be done to prepare the soldier for the day when his equipment does arrive. First, and most important, to provide a general education in the basic concepts of data processing and the many ways in which it can be used as a tool; the advent of the microprocessor has made the field of application almost boundless.

> "The application of microprocessors is simply limited by the ingenuity of the human mind".

Second, in the case of C^3 systems, to educate those involved to a greater depth in communications technology and the factors involved in command control, and decision making. Third, to ensure that as a result of the knowledge and, hopefully, the enthusiasm achieved in the first two stages, the soldier does and is keen to have an input to the design and development of his equipment. User input is vital to the successful implementation of any ADP project but such input is markedly more valid if it comes from individuals with a grasp of the basic technologies and possible applications; this means 'educated users'.

Digital Transmission

Digital transmission itself presents a worry to many new military users. It has been proved that data is a much more efficient method of passing information than either voice or a mixture of voice and data. Data transmissions take microseconds and will thus take less time than voice transmissions; nevertheless there may well be a temptation to increase the amount of data transmitted and the number of messages to fill the space now available, thus aggravating even more the problems of information handling. C^3 systems will probably therefore need to impose a selective priority constraint on all messages. A switch to data only transmissions on the battlefield and do away with voice, will create some contentious human problems. It has been suggested, for instance, that there could be a loss of the ability to create a sense of urgency or to dispel the grave fear of feeling lonely. There is also a fairly common feeling that a generation of military managers is being produced, as opposed to leaders. Such fears must be balanced against the realisation that battlefield command and control will probably depend on digital data transmission for it to work at all: it is easier to provide a variety of interchangeable communications modes on data-only nets, thus providing a better capability to work through jamming.

Training Aids

It is possible to build training aids for operators and users of ADP equipment. It is also possible to build computerised training aids to meet many other types of training requirement. This whole field is developing rapidly but could prove expensive if the requirements are not clear and strictly controlled. The danger is that many parallel courses could evolve with a tremendous waste of development skill and duplication of effort. It would be much better if common software could be written. ADP support can be divided into training in individual skills and training in tactics. Let us consider three areas where aids to improve individual skills are being developed. First is the training of tank crews which is limited by the availability of fuel, ammunition and training areas. Simulation techniques are already widely used with a variety of levels of sophistication, from those which allow very realistic training to be carried out within the tank turret to those employing purpose built indoor simulation systems. Training systems which interact with other weapons have been developed; they consist often of a laser unit mounted on the gun barrel, a control unit, transmitter receiver, flash generator and pyrotechnics for noise and smoke. After selecting a target to engage, the loader sets the range and ammunition type on the control panel. The gun is fired after a pause to simulate the time taken to load, and a series of laser pulses are transmitted; at the same time a flash is generated. The laser pulses are detected by sensors on the target and the exact range is measured to compare with the gunner's estimate. The control unit calculates fall of shot, taking account of the ammunition type, and a lamp in the sight indicates a hit or miss. It is a simple but effective aid.

Fig. 8.1 Tank Gunnery Training Theatre

Figure 8.1 shows equipment developed for the basic training of tank gunners using the IFCS (Improved Fire Control System) system produced by MSDS Ltd.

In comparison, tank crew trainers to provide skill training for tank commanders and gunners are available. They consist of replica turrets, tabletop terrain models and fairly large computers to control all the data. The replica turrets are completely fitted with standard operational equipment less, perhaps, the laying and stabilisation hydraulics. The instructor occupies the position of driver/loader and has a console showing the view through the gunner's sight and readings of all the variable parameters such as barrel wear and weather conditions. He is able to input problems such as fog or equipment malfunctions. The trainees have a view of the terrain model, produced by CCTV, which is supplemented by a variety of mobile and fixed targets; these models can be equipped with flash simulators.

When an exercise is in progress the gunner has a changing scenario and apparent movement of the tank augmented by realistic mechanical noises. Once a target is acquired the gun is fired in the normal way, accompanied by realistic vibration and noise. The firing data from the fire control system is matched with information already held in the computer and the shell trajectory is computed and displayed, with the shell appearing in the sight as a luminous dot. The computer follows the trajectory in a good deal of detail and will record a target, or ground, impact. A target impact can be signified by a flash or explosion. Information on each shoot is stored by the computer for later analysis.

Fig. 8.2 The same view for gunner and instructor

Figure 8.2 shows the gunner's view through the IFCS sight which is also reproduced for the instructor in the training system shown in Fig. 8.1.

Artillery has similar training constraints to those of armour but the training of an artillery observer is even more difficult in view of the large amount of ammunition required. Several artillery observer trainer systems have already been in use for some years. Some consist of a large panoramic display, produced by colour slide projectors. The display can be supplemented by target symbols or images as part of a TV projection or, as in some more modern systems, by means of optical servo systems which provide clearer images. Most types of fire mission can be simulated and the effect of the artillery fire can be augmented to include smoke, shrapnel and noise. Exercises are normally set up at a keyboard and VDU console where the composition of the artillery units is decided and various data such as meteorological corrections, ammunition types, and times of flight are input. Exercises can often be recorded for playback and analysis. The range of facilities in this area is large and development is still continuing.

Aids for anti-tank missile training are in a fairly early stage of development and consist, most commonly, of the projection of a luminous dot into the sight to represent a tracking flare. Luminous symbols to represent moving or static targets can also be projected into the sight and a computer used to compare tracking with known data. The next stage can be to provide a monitor sight for the instructor for it is only by such close involvement that proper advice can be given.

Despite the fact that there is some involvement by individuals other than the trainee in each of these systems it is really only the trainee who is being trained in his manipulative skill. In no way do commanders get any experience of the tactical use of the weapons being simulated. To do this it is necessary to re-create an environment with all its influences. Air forces and navies are already well aware of the problem and have gone a long way to providing suitable trainers. For example several naval instructional establishments have simulations of ship operations rooms. Such systems do, however, have two distinct advantages over their army counterparts: first, most of the inputs to such an operations room are provided digitally, from radar, sonar, radio or other communications links, and these are easier to simulate on identical displays to those in current service. The army commander, at present, has few automatic digital inputs and relies heavily on verbal dialogue with other cells; this is difficult to simulate because many staff and much equipment may be required to produce an interactive environment. Second, the army commander, at the lower level, has a requirement to see what is going on and is often remote from his command post. Providing aids to help the training of the commander at a low level has proved difficult; but at the levels where military commanders first start to make decisions based on data input to a cell where it is co-ordinated, rather than on a physical view, there has been some progress. With the advent of ADP C^3 systems and digital inputs these higher level cells will also gradually prove easier to simulate; but the cost of software to provide fully automated, integrated, interactive training systems should not be underestimated. There may be room for partial automation only, particularly as experience shows that both players and staff can receive almost equal benefit in such systems.

Pattern Recognition

Target acquisition and remote surveillance are important military applications of pattern recognition. At the Royal Military College of Science research work is in progress on the application of syntactic pattern recognition to these problems.

Syntactic pattern recognition is based on formal language theory and has close links with the methods used to construct compilers. In principle, a set of reference patterns is summarised by using the sequences of measurements obtained from them. An unknown pattern is classified by comparing the sequence of measurements obtained from it. A simple example would be the recognition of the outline of a tank because the sequence of steps around its outline matches that of one of the stored reference patterns. This sequence will differ from that obtained from the outline of, for instance, an aircraft. The method can only be effective if the summary is sufficiently general, and approximate comparison is used.

There is evidence that no single type of surveillance sensor can provide adequate performance under poor conditions of signal to noise ratio in the field. Effective use of more than one sensor requires the recognition algorithm to perform 'information fusion' of signals from, for example, TV and IR sensor systems. Syntactic pattern recognition may provide a formal basis for such algorithms using a mathematical device known as affix grammar. At its simplest, this tags rules in the grammar with additional facts about their properties. A fired tank gun or its engine cowl will be a stronger infra-red emitter than the rest of the hull. A strong straight line in the TV image of the same infra-red emitter is more likely to be the gun than the side of the hull. In this way the information from the infra-red sensor can help the recogniser label the TV image.

It is expected that the outputs of other types of sensor may also be combined in this way and the resulting recognition should show only a graceful degradation in performance as some of the inputs are removed due to weather or enemy action. Where a particular pattern is expected, the image can also be enhanced by exaggerating features.

Voice Synthesis and Recognition

Since the mid 1970s a good deal of work has been carried out to develop voice synthesis technology. There is now a broad range of products from voice synthesiser chips and specialised voice memories, to speech development laboratories which allow the development of individual electronic speech in much the same way as the writing of a microprocessor program. Pitch, energy and vocal tract filter information is extracted from human speech recordings to produce 'speech instruction data' which can be stored at typically 100k bits per 1 second of speech. This data can then be used, with a voice synthesis processor, to recreate the human voice. 'Custom' words and phrases can be generated for particular applications and users would be required to record their voices using each of the words or phrases in order that the computer can recognise them at a later date. Voice recognition has great potential, but has yet to be proved suitable for military use. It has the constraints that the user's voice must be within the voice capability of the processor and that the user's voice should remain at a standard pitch and energy: this might prove difficult in battlefield conditions.

Management of In-Service Software

In Chapter 4 the process of configuration control in development was described and it is emphasised that in no military equipment procurement area is there more need for such a mechanism than in that of software for ADP equipment. It does not matter whether the equipment is part of a large system or an application nested within another larger program. In ADP equipment there are almost invariably interrelationships between programs, and programs are necessarily run concurrently or in parallel to avoid time wasting and improve performance. Thus any change to a requirement, however small, may have widespread and possibly dramatic implications in many other areas. Strict control must be exercised on the overall software plan and changes must be managed by very careful configuration control. During development changes may be caused by:

Changes in the requirement.
Shortage or delay of funding.
Inability of the manufacturer to carry out the necessary development.
The unknown.

The proposed changes will normally be assessed by the manufacturer and will then be presented to the project manager who will balance cost/time penalties against the change and present his recommendations to the sponsor and user. Changes that are agreed by all parties will be implemented and the results should be reflected in all the affected work areas, including test plans.

To achieve effective configuration control there must exist a baseline of accurate documentation which details both requirements and design. This is also very necessary to manage the project efficiently against a contract. Loose, inadequate documentation will lead to loose contractual arrangements and complications on cost and completion dates.

It is clear, therefore, that the task of configuration control during development, even with experienced project managers, professional financial advice, dedicated manufacturers and knowledgeable technical authorities, is very time consuming but vitally important.

Having achieved a plan for managing software, and changes to it, during development, each piece of software is written and should be painstakingly verified, validated, checked for quality and then tested against predetermined requirements in order to assess its acceptability. The equipment is then built and issued to the user who may wish to carry out more tests and trials but by this stage has paid for it. How can the user now tackle the problem of software changes and maintenance. To emphasise the magnitude of the maintenance task we saw in Fig. 3.8 that, during software development, analysis and design formed 40% of the task, testing 40%, and coding and debugging 20%. Now consider Fig. 8.3 which shows that, in the total life cycle of a computer system, maintenance forms some 70% of the total software task.

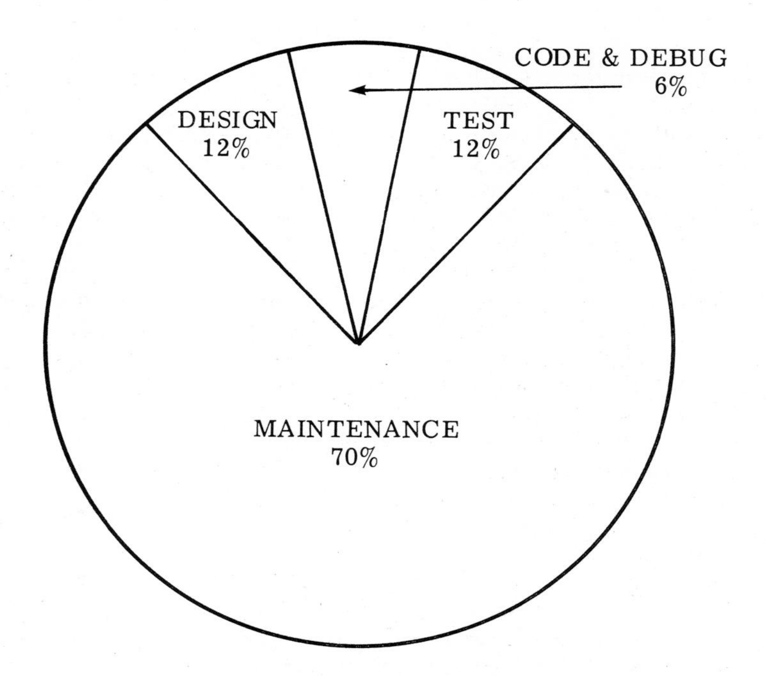

Fig. 8.3 Maintenance in the life cycle

There are usually two tasks to be carried out: first, debugging which, as we know, is the process of correcting errors in software which are not discovered during the test phases of development. It is a common problem due mainly to the invisible nature of software; it is not practical to hope to design a test plan which caters for all possible manipulations of software.

Then there are amendments and enhancements. These arise because the life cycle of an ADP system will, in many cases, never be completed. For instance there is normally an ever changing and poorly defined requirement. Very often a prototype approach is necessary to allow the user to familiarise himself with ADP and its capabilities before he is capable of specifying his requirement in detail. It is possible to build in flexibility to allow for this by providing sufficient 'hooks' on which extra peripherals and store can be hung, but there will be more software to be written and the cost of this could be very difficult to estimate.

The military user has a choice of three solutions to his software maintenance problem now that he owns the computer. First he can continue to employ the manufacturer who provided the software to maintain it during its service life. This approach has, at first sight, obvious attractions of continuity but could have some serious drawbacks. The manufacturer's team could be very expensive to hire over many years and is unlikely to be acceptable on the battlefield during hostilities if, indeed, the team wishes to stay. Then if the particular item of equipment did not have further sales potential the manufacturer may not be too keen to tie up key staff to support the equipment for an indeterminate period.

The second choice is to train military staff to carry out the task. This could have enormous advantages. The main one concerns close control and thus, theoretically, better management. This approach also provides a highly trained military pool of staff to specify the requirements for new systems. It also overcomes all the problems of the manufacturer team approach but it has its own drawbacks. The army in question may not be sufficiently large to provide the manpower cover. The soldiers would probably not be cheaper than manufacturers' employees when quartering, food and administration costs are considered. It also may not be possible to train the soldiers to a high enough level of expertise to carry out the task. If they could be it would be difficult to prevent them leaving the army to join manufacturers who may pay higher salaries.

The third choice is a level of compromise between the first two choices. It may prove sensible to train military personnel to a level where they can identify problems and suggest sensible and feasible enhancements which could be implemented by a small, efficient manufacturer team. The final choice may well depend on the complexity of the system under consideration. It may be logical to give most user-oriented tasks to military teams where it is considered that a military team is viable and more complex tasks to supplementary staff from industry.

Interoperability

In order to promote the ideology of 'system thinking' it is vital that equipments which form parts of an ADP system should be interoperable. This problem must be addressed right from the concept stage for the parts may be from different manufacturers, from different services, and from different nations. Interoperability is the ability of automated systems to exchange and use information; there are three facets of interoperability which must be considered together: they are degree, standards, and levels. Consider Fig. 8.4.

Degree 1 is, in fact, two manual interfaces; Degree 2 is the use of two liaison detachments with remote terminals from the other system; Degree 3 is an enhancement of Degree 2 but with some automation such as magnetic tape to transfer data at the interface; Degree 4 is automatic but with a measure of control on the free exchange of data and Degree 5 is automatic with no constraints which make the two terminals appear to be part of the same distributed system. Degree 4 appears to be the type of mechanism which might be expected to be set up between two nations. There would be agreement on the majority of data to be passed between the two; but national privacy or security caveats would cause some data to be held back for, at least, consideration before transmission.

Standards for interoperability can be logically divided into three groups. Operational standards consist of agreed military objectives relating to plans for exchange of specific information. Then there are procedural standards which concern the form in which information is transferred and how it will be handled. Finally technical standards are required for functional electrical and physical characteristics to allow the transfer of information between different items of equipment.

As far as levels of interoperability are concerned it is also possible to divide these into three; in this case they are international, national and system. International interoperability between friendly nations will usually be arranged formally in an international forum such as NATO. Nationally there will be a need for communications and other ADP projects to decide on interoperability plans at an early stage to avoid expense later. At a system level, system components should be built to common standards. At all levels there will be much discussion of the type and quantity of data to be transferred between cells; it is important that a firm plan is made, for this identification of the information to be transferred is a measure of the requirement for interoperability.

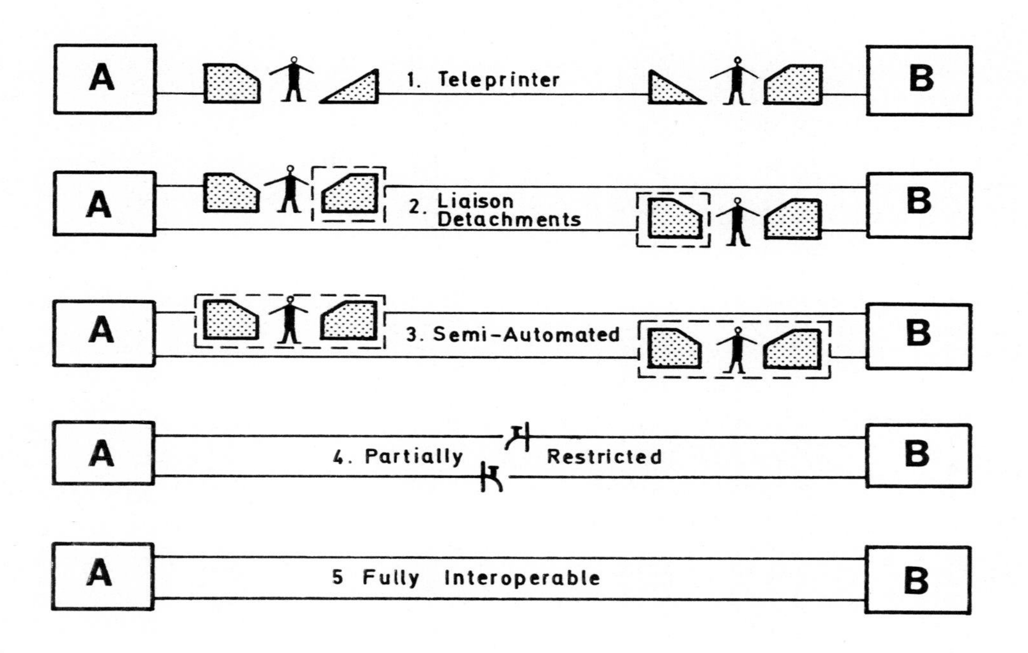

Fig. 8.4 Degrees of interoperability between ADP systems

The importance of a consolidated approach to interoperability and the management of the interfaces between equipment cannot be stressed too much. Failure to plan ahead could prove to be very expensive at a later stage when interoperability requirements are uncovered.

"Nine tenths of wisdom is being wise in time"

Theodore Roosevelt.

SYSTEM EVOLUTION

The Computer Industry

The early advances in military computer technology were, naturally enough, engineered by large electronics manufacturing firms who had proven skills in the production of circuitry and in the installation of electronic equipment. There is no doubt that those firms produce excellent hardware but due to their nature they have not, in many cases, managed to provide the kind of system and software expertise that the users of larger, modern systems require. The same problem has occurred in the larger civilian market and the first sign of a change in the industry was a movement to 'software houses'. They specialised in grouping software experts together and in being sub-contractors to the bigger 'hardware oriented' firms for specific projects. The next development was to 'system houses': some software houses have built strong design teams who offer to analyse the system, design a solution, develop the software, and suggest a preference from a variety of hardware devices on which the software could be run. This is perhaps, a very useful service for a customer who does not want the task of selecting between hardware firms.

For smaller systems a plethora of good quality manufacturers exist; most have the sensible attitude that "the computer has to adapt to the user, rather than the other way round". This quotation is from Steven Jobs, whose Apple Computers sales progressed from £1.5M in 1977 to £65M in 1980. With the advent of a knowledge based society there is a requirement for people to be trained in the evolving technologies and the computer is one of the first tools to amplify intellectual ability rather than inherent physical ability. The older 'giants' of the computer industry have in many cases failed to keep pace with the growth of the market. As a result, large numbers of smaller manufacturers, with their more personalised approach to education, MMI, and quick service, have capitalised on the advent of the very reliable microprocessor.

The Semiconductor Industry

Computers use large numbers of electronic components: consequently there has always been some tendency for semiconductor manufacturers to set up specialist computer manufacturing divisions to provide a captive market for their components. These divisions, which have not always been successful, have used their parent company's SSI and MSI devices, but since so much specialist design effort was required to produce a computer from these building blocks the main company and its computer division had little technical impact upon each other. A slightly different situation exists in some of the mainframe and minicomputer companies where divisions have been set up to manufacture SSI, MSI, LSI or VLSI devices designed specifically for use in their computers. Nevertheless, despite this blurring of the edges there has been a general tendency in the past for computer manufacture and semiconductor fabrication to exist independently, even to some extent in isolation from each other.

The introduction of LSI and VLSI devices, such as the microprocessor, has changed this situation significantly. An MPU consists of the CPU of a computer in one integrated circuit, so this means that the semiconductor manufacturers have become intimately involved in computer design. In the early 1970s MPUs were designed without too much regard for some of the lessons learned by mini and mainframe computer manufacturers over the preceding decades. It is sometimes said that the early MPUs put computing science back a decade in terms of their design as computers. To some extent this was because it was only just possible to fabricate a simple CPU on a single silicon chip; but it was also true that the design was being carried out by experts in semiconductor technology and not computing. However with the trend towards an ever larger number of devices per chip the semiconductor manufacturers have been able to build quite powerful and sophisticated MPUs; the expertise gained from the earlier devices, coupled with the setting up of design teams with specific computing backgrounds, has enabled them to advance in this way.

The corollary of this expansion of MPUs into the area occupied by minicomputers is that semiconductor manufacturers are moving from being purely component suppliers to being component and system suppliers, because they can readily use their components to build complete computers. These systems may well undercut in price those from conventional minicomputer manufacturers who do not have access to LSI and VLSI design and fabrication facilities. Although software support from the semiconductor industry has tended to be poor, it is improving: consequently the next decade is likely to see dramatic changes in the structure and composition of the minicomputer manufacturing industry.

Continuous System Evolution

The implementation of Military ADP systems follows the pattern of most military equipment which is based on a development model which assumes a life cycle. This is comprised of a number of discrete development stages, followed by a predicted but finite operational life, which ends with its replacement by a new system. This pattern assumes that it is possible to analyse the requirement, to base the design on the analysis, and to implement the design in a time frame normally measured in years. There are a number of well known flaws in such assumptions. When the system is eventually implemented it reflects and meets the requirements as stated some years earlier and not those existing currently. Then the system will be based on a set of ideal designs to meet ideal requirements; but it is very difficult to know how the user will react to the design or whether it is really what the user wanted and to test it out at intervals is normally impossible. Finally, the development process is carried out by professional ADP experts. Normally they are not military and work remotely from the eventual user. As a result the developers fail to gain the required depth of knowledge about the system that a user would possess.

Recently several fresh approaches have been discussed and tried. First, as mentioned several times earlier, continuous and early user involvement should be encouraged to allow the user to share in the design of the system. Tools to aid definition, to diagnose system shortcomings, and to help evaluate alternative design will be necessary but it is incumbent upon the user, through his procurement agency, to demand such facilities.

A second approach is the concept of continuous system evolution rather than a life cycle. Any changes would be considered as enhancements or increments: they would tend to be fairly small providing they were implemented as and when necessary as the result of regular review. Such a concept depends upon the correct amount of flexibility, expandability or redundancy being built into the design from concept. Several technological advances have made such an approach more acceptable. Modular and structured programming techniques and database management, which are by their very nature dealing in compartmentalised, stand alone modules, lend themselves readily to expansion.

The use of stand alone mini and micro systems can also be looked upon as a useful tool in this context where they can be employed as localised, dedicated systems and possibly locally controlled. There is, however, a grave danger that development could proceed in an unco-ordinated fashion, often being undertaken by insufficiently skilled staff. Many tasks could be duplicated or, on the other hand, mismatched. The answer is the provision of an efficient management system.

The third approach is to recognise that the designer's ability to predict the behaviour of an ADP system, and those who are going to use it, is poor. Design based on what is assumed to be a finite analysis of a requirement, supplemented by statements of what is expected to be a behaviour pattern for the user, is likely to be less than fully effective. To overcome this, experimental techniques must be introduced into the design process. The most common, and proven in the case of WAVELL, is the prototype approach. Although apparently inefficient in its use of resources, it allows the designer to test the validity of his assumptions. It also allows the user to take his first steps into the unknown and to embark on the first part of his practical learning curve; as a result of this the user can be expected to provide his first objective input to the project.

CONCLUSION

We have considered what computers are, how they work, what some of their military applications are and what advances there are on the horizon. We have also stressed several themes throughout the book.

User involvement at all stages of a computer's life, from concept to in-service maintenance and operation, is essential if the computer is to be used to its full potential and if the requirement is to be fully met; but the user representation must not be confined to technical arms, for the user will only be fully representative if he is experienced in the area which is to receive computer assistance.

As with all other military projects it is essential, in order to provide the optimum cost effective solution, to specify very clearly what the task, or requirement is. We have seen that, due to the nature of computing and the lack of understanding of this new technology, and despite the start that has been made in educating users, adequate specification of the requirement is seldom achieved. A prototype approach, in which the user is given a sample of a tool that he might find useful

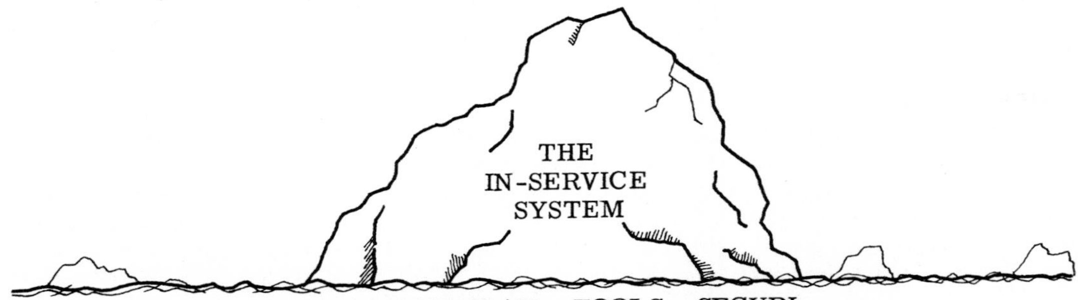

SOFTWARE PLAN. TOOLS. SECURI
TY. HARDWARE SELECTION. TRAIN
ING. BITE. QUALITY ASSURANCE. AT
E. MMI. TRIALS. EXPANDABILITY. TE
STING. REAL-TIME. ENCRYPTION. LANGUA
GES. COMPILERS. EXECUTIVE. VOLATILITY.
INTEROPERABILITY. FEASIBILITY. MODULARI
TY. COMMUNICATIONS PLAN. SYSTEMS ANALY
SIS. ARCHITECTURE. CAPACITY. DEBUGGING.
ADDRESSING. ASSEMBLERS. REQUIREMENTS.
SPECIFICATIONS. PERIPHERALS. COMPATA
BILITY. USER INVOLVEMENT. MULTIPROGR
AMMING. ACCESS TIME. HOST/TARGET.
OPERATING SYSTEM. ON/OFF LINE. S
ECOND SOURCING. SUPERVISOR. SP
EED. DATABASE. DISTRIBUTION.
DATA. INFORMATION. NUCLEAR
HARDENING. PROGRAMS. RA
M. MODULAR DESIGN. ST
RUCTURED PROGRAMMI
NG ETC.....

Fig. 8.5 The finished product

and is permitted to experiment with it to allow both the true requirement to evolve and him to gain experience, has proved its worth.

For several reasons, software is far more expensive than hardware to develop. There can be no error tolerance in software but it is invisible and in development it is difficult to size, cost, test and monitor. Costs can escalate unless strict rules are imposed and tools are used to aid management. Almost without exception systems are more likely to meet their requirements if the top level software design is completed before any decisions are made on the specification of hardware to support the design. Hardware is comparatively inexpensive and is in plentiful supply; in most cases there will be a selection of several suitable components from which to choose.

It is worth considering the total amount of interrelated effort which is necessary to bring any but the simplest battlefield computer into service. It can be seen in the iceberg shown in Fig. 8.5. Seldom can military computer systems be purchased "off-the-shelf" and all the effort shown below the waterline will be needed to achieve the apparently smaller product which sees the light of day.

Finally, we have discussed 'systems' and 'system thinking'. In computing on the battlefield there are five key areas where it is vital, for both operational and cost reasons, that high level management directs development to achieve effective systems; modularity, commonality, expandability, interoperability and the correct sequencing and phasing of projects coming into service.

Answers to Self Test Questions

CHAPTER 1

Page 21

QUESTION 1 Calculation and printing out of previously incomplete mathematical tables.

QUESTION 2 In an analogue computer numbers are represented by the magnitude of some measurable quantity (eg an electric voltage). In a digital computer an individual indicator is used for each digit.

QUESTION 3 A system is a group of related or interacting parts forming a unit. All systems form part of a larger system.

QUESTION 4 A BIT is a binary digit, a 0 or a 1, and is the basic unit of information used within the computer.

QUESTION 5 A BYTE is a conveniently sized collection of BITS, normally 8, for moving data around inside a computer but does not necessarily represent a complete unit of information.

QUESTION 6 A WORD is a unit of information and normally consists of 16, 24 or 32 BITS.

QUESTION 7 Arithmetic and Logic Unit, Main Store, and Control Unit.

QUESTION 8 Direct access implies almost immediate access of a data file, normally achieved by having several points of access to the store. Serial access implies the act of having to read each item of information in a store serially until the item required is identified, as in a magnetic tape passing reading heads. The effective difference is in access time.

QUESTION 9 A program is a set of instructions or rules which govern the activities of the various parts of the computer.

QUESTION 10 To interpret and execute the individual program instructions by carrying them out as a sequence of tasks and initiating the appropriate operations by the other computer components.

CHAPTER 2

Page 44

QUESTION 1 'Volatile' means that data is lost when power is switched off. (Semiconductor RAM is normally volatile).

QUESTION 2 ROM (Read-only memory) - data can be read from it but cannot be written to it. Used when it is essential that a program or data should not be altered.

RAM (Random-access memory) - data can be both read from it and written to it.

QUESTION 3 Access time, capacity, volatility, cost, physical characteristics, security, availability, standardisation.

QUESTION 4 8 bits is the word length and is the maximum width of the registers within the ALU. The CPU therefore handles data 8 bits at a time and will have an 8 bit data bus. In general terms the higher the figure (eg 16 and 32) the greater the power of the CPU. For scientific work with floating point numbers a long word length is desirable, however many data processing applications are concerned merely with the manipulation of characters and a shorter word length may be adequate.

QUESTION 5 a, d, b, c, e. In general the increase in access time is proportional to the decrease in cost per bit of store.

QUESTION 6 Small size. Non-mechanical and therefore no moving parts. High capacity. Low cost compared with semiconductor.

QUESTION 7 Core store is non-volatile.

CHAPTER 3

Page 81

QUESTION 1 Low Level Languages are those close to machine code which can be recognised and used directly by the machine. They are normally specific to particular machines and the purpose of a low level program is seldom evident to anybody other than the individual programmer.

QUESTION 2 High Level Languages are designed so that each instruction corresponds to several machine code instructions and each instruction can be written in familiar English words and common notations.

QUESTION 3 A compiler translates single source (high level) statements or instructions into several machine instructions.

QUESTION 4 High Level Languages are relatively inefficient in their use of store and do suffer some lack of flexibility due to the limited number of instructions available in a standardised language. They are slow to run.

QUESTION 5 The capability to run any program on any machine.

QUESTION 6 Designing-to-test must be encouraged to ensure that software meets the requirement. The test must reflect the requirement and should be defined before any programs are written. This will allow the testing of software in modules as it is produced.

QUESTION 7 The ability to run programs apparently simultaneously is essential if optimum use of all peripherals and thus maximum processing speed and efficiency is to be achieved.

QUESTION 8 An assembler is used to translate assembly code into an object program or machine code.

QUESTION 9 a. (1) A large part of design effort and development cost is now spent on software rather than hardware.

(2) Software is 'invisible' and its development is difficult to monitor.

(3) Software tools can be used to impose rules in this potentially poorly organised area and will thus provide elements of control.

(4) The design, implementation and maintenance of software demand management; good management can only be achieved by setting clear rules and having visibility of progress.

b. 'Aids' could include:

(1) Well supported high level languages.

(2) Prototyping - host/target.

(3) Enforcement of standards.

(4) MASCOT.

(5) Development of good programming techniques.

(6) Top-down design.

(7) Structured programming.

(8) Systematic design reviews.

(9) Computer assistance such as SDS.

(10) Well planned, logical, structured testing.

'Tools' are perhaps (4) and (9).

QUESTION 10 a. Modular design is the technique of analysing the task which is being performed into a hierarchy of sub-tasks. To each sub-task is assigned a module of code which has a well defined interface with the remainder of the hierarchy.

b. Structured programming is a technique designed to lay down a set of guidelines for the writing of 'clean', 'gimmick-free' programs using the modular design concept. It involves, first, teaching programmers to program rather than specific programming languages; second, getting the program correct by systematic design rather than trial and error.

CHAPTER 4

Page 100

QUESTION 1 The methodical investigation or examination of a problem and the separation of the problem into smaller related units for further study to decide whether ADP is necessary; if so to what extent and how it should be implemented.

QUESTION 2 A flowchart is a pictorial method of presenting a process in a logical progression of steps.

QUESTION 3

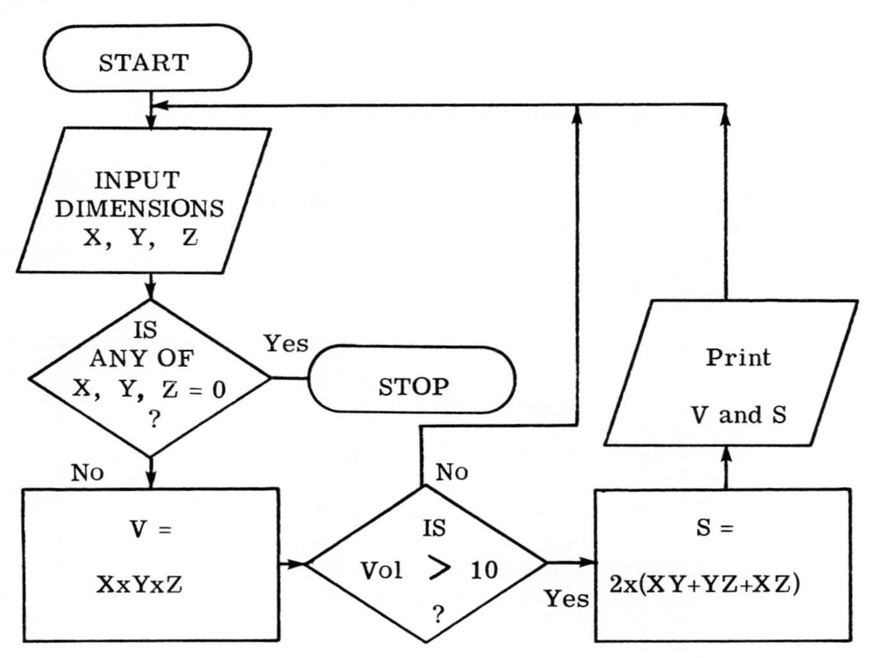

QUESTION 4 Top-down design is the process of designing iteratively at increasing levels of detail starting at the top.

QUESTION 5 The preliminary study (often termed 'feasibility study') is used to decide whether or not ADP is necessary or desirable. The preliminary study report is used as a basis to determine whether the project should proceed.

QUESTION 6 First, that education of the new user will take time. It may be sensible to introduce a simple system first or the 'ideal' system by stages. Second that the 'ideal' solution will contain untried techniques and thus an element of risk; such risks must offer significant benefit to be worthwhile.

CHAPTER 5

Page 123

QUESTION 1 The microprocessor corresponds to the CPU.

QUESTION 2 A silicon chip is a segment of a silicon wafer which contains transistors and other components.

QUESTION 3 The size of the silicon wafer and the number of connections necessary dictate the size of the package.

QUESTION 4 2^{20} or 1,048,576 locations can be accessed directly by a 20 bit address bus.

QUESTION 5 Bit-slices are used in order to obtain improved speed.

QUESTION 6 Benefits: only one package therefore very small size of equipment and low manufacturing cost are possible.

Limitations: the amount of memory and I/O connections available are limited.

QUESTION 7 The software is specific to each design, and so requires development effort, whilst the majority of the hardware is standardised.

QUESTION 8 Burn-in is used to identify those circuits which will fail during the infant mortality period and so prevent their use in production.

QUESTION 9 The vast range of current and possible applications makes it impractical to attempt to catalogue them here. References to microprocessor applications will be found not only in specialist journals but also in daily newspapers, motoring magazines, and so on.

CHAPTER 6

Page 135

QUESTION 1
a. Limit the numbers of people involved.
b. Define what level of access they require.
c. Provide coded doorlocks, sentries and physical barriers.
d. Enforce personal identification.
e. Shield against electromagnetic radiation.

QUESTION 2
a. To verify that access to a store to carry out an instruction does not contravene a prearranged access restriction.
b. To authorise access once it has carried out the verification.

QUESTION 3 To check that design of a program meets the intent of the specification.

QUESTION 4
a. Logging in of operators.
b. Frequent authentication of operators during operation.
c. Restriction of access to data other than where essential.
d. Encryption of all data transmissions.

CHAPTER 7

Page 174

QUESTION 1
a. Nuclear hardening - hard devices are essential.
b. Vibration, temperature, environment etc.
c. Continuity of supply - second sourcing.
d. Commonality and modularity make maintenance and training easier and provide good fall-back capabilities.

QUESTION 2 A real-time system is a system which controls an environment by receiving data, processing it, and returning or outputting the results quickly enough to affect the functioning of the environment at that time. Important points are:

a. Speed is essential.
b. Communications may be a larger problem than processing.
c. Multiprocessing will be essential.
d. The necessary fast access store will be expensive.
e. MMI will be most important - the machine must be 'fitted around' the operator.

QUESTION 3 BITE consists of software designed to test hardware in order to detect errors or failures as soon as possible after they occur.

QUESTION 4
a. FACE employs bit-slice microprocessor circuits for ballistic computation and the display.

b. BATES employs a multimicroprocessor configuration to handle ballistics, communications and display. The use of a number of microprocessors enables speed to be achieved whenever desirable.

QUESTION 5 a. The user helps to define the requirement as a result of experience with the prototype.
b. The user gets equipment earlier, albeit prototypes.

QUESTION 6 Processing power is sited at each module. A "distributed database" implies that each cell holds a complete duplicate database.

QUESTION 7 a. No manual intervention.
b. E, D and C techniques check the accuracy.

QUESTION 8 Emulation is the process of using a computer to operate on data and programs originally produced for a different computer; special software and hardware are utilised to represent the original computer.

QUESTION 9 Resilience; work can be switched away from a faulty processor and distributed to other processors, thus allowing the multiprocessor configuration to remain in operation, albeit in a slightly degraded mode.

Glossary of Terms and Abbreviations

A

Access Time
: The time that elapses between giving an instruction to access a storage location and the moment when the transfer of data begins.

Address
: The reference number which uniquely identifies an area of computer storage.

ADP
: Automatic Data Processing.

ALGOL
: An internationally agreed computer language for scientific use. ALGOrithmic Language, primarily used to express computer programs by algorithms.

Algorithm
: A sequence of rules defining a computational process.

Analogue Computer
: A computer which processes numeric information represented by physical quantities such as voltages or rotation of shafts and gears.

APSE
: Ada Programming Support Environment.

Architecture
: System - The fashion in which hardware and software fit and communicate together.
Hardware - Design of the CPU.

Arithmetic and Logic Unit (ALU)
: Part of the CPU which performs the necessary operations on the data.

Assembler

Software program which translates instructions in assembly code to binary machine language.

Automatic Test Equipment (ATE)

Diagnostic equipment used in the manufacture and maintenance of ADP systems.

Automatic Data Processing (ADP)

The rapid manipulation of data by electronic or mechanical means. Hence ADP System (ADPS).

B

Backing Store

The secondary storage level of a computer. It is cheaper to construct than the primary level. Examples are magnetic drum, magnetic disc and magnetic tape.

Batch Processing

A system in which data is accumulated to form a batch, the batches then being processed at regular intervals.

Binary

A number system based on powers of 2. The digits, or bits, are '0' and '1', and may conveniently be represented by a circuit being turned 'on' or 'off' or a magnet being polarised '+' or '-'.

Binary Digit (Bit)

The smallest unit of information consisting of a single value '0' or '1' in a binary number. Contracted to 'bit' from BInary DigiT.

Built-in Test Equipment (BITE)

Diagnostic software and/or hardware built in to a computer.

Bus

A parallel set of connections linking parts of a computer.

Byte

A group of 8 bits.

C

CAD

Computer Aided Design

CAL

Computer Aided Learning.

CAM
Content Addressable Memory (or sometimes Computer Aided Manufacturing).

Capacity
Used in conjunction with storage to describe the number of computer bytes or bits that can be held within the store.

Cathode Ray Tube (CRT)
Common device used to produce a television-like display. Used in VDUs.

CCTV
Closed Circuit Television.

Central Processor Unit (CPU)
The main part of the computer consisting of the arithmetic and logic unit, the control unit, and sometimes the main store.

Chip
A piece of a silicon wafer which contains the transistors, resistors and capacitors forming an integrated circuit.

CMOS
Complementary Metal Oxide Semiconductor (uses both positive and negative charges to transmit data).

COBOL
A computer language designed for commercial purposes (COmmon Business Oriented Language).

Compiler
A special program which acts as a translator between a program written in a high level language and the machine code.

CORAL
A computer language designed principally for real-time operating applications. It is now the UK standard language for defence work. (Common Real-Time Applications Language).

Core Store
A storage medium based on the magnetising of small ferrite rings.

CPU
Central Processor Unit

D

Databus
A group of parallel connections used to carry instructions and data between computer components.

Debug

To fault-find in an erroneous program.

Digital Computer

A computer which operates by holding discrete numbers (normally in binary form) rather than by measurement of some continuously varying quantity.

DMAC

Direct Memory Access Controller.

DRAM

Dynamic Random Access Memory.

E

EAROM

Electrically Alterable Read Only Memory.

EEPROM (or E^2PROM)

Electrically Erasable Programmable Read Only Memory. (Another name for EAROM).

EPROM

Erasable Programmable Read Only Memory.

F

File

A collection of related data treated as a complete unit within a computer system.

Firmware

Data or instructions stored permanently in ROM.

Flow Chart

A conventional technique to show the logical sequence of steps which makes up a particular process. Used in systems analysis and program writing.

FORTRAN

One of the earliest computer languages designed by IBM for scientific and engineering applications. FORmula TRANslator, the mathematical equivalent of COBOL.

H

Hardware
The physical pieces of equipment which make up a computer system.

High Level Language (HLL)
A programming language which nearly equates to ordinary language or mathematical notation.

I

IC
Integrated Circuit.

I/F
Interface.

I^2L
Integrated (or Current) Injection Logic.

Interrupt
A facility which allows the suspension of a current program while an alternative 'Interrupt Handler' program is executed.

I/O
Input/Output.

IR
Infra-red.

K

K
Approximately 1000. Actually 2^{10}=1024. Used as a shorthand way of describing storage capacity, etc.

L

LAN
Local Area Network.

LSI
Large Scale Integration.

M

M
Approximately 1,000,000 and actually 2^{20}= 1,048,576.

Machine Language (or Code)
A language that is used directly by a machine consisting of instructions etc in binary code.

MASCOT
Modular Approach to Software Construction, Operation and Test. A technique for software production and management.

MCU
Microcomputer Unit.

Memory
Synonymous with computer store, holding data and programs.

MICR
Magnetic Ink Character Recognition.

Microprocessor
A very small central processor based on large scale integrated circuit (LSI) technology.

Microsecond (μs)
One one-millionth of a second.

MMI
Man/Machine Interface.

Mnemonic
An easily remembered substitute for a word or phrase.

Modem
A device that modulates and demodulates digital data signals transmitted over communications facilities.

MOS
Metal Oxide Semiconductor.

MPU
Microprocessor Unit.

MSI
Medium Scale Integration.

MTBF
Mean Time Between Failure.

MTTR
Mean Time to Repair.

Multiprogramming
The technique of operating 2 or more programs in a computer at the same time.

N

Nanosecond (ns)
One thousand-millionth of a second.

NMOS
N channel Metal Oxide Semiconductor (uses negative charges to transmit data).

O

OCR
Optical Character Recognition.

Off-line
Pertaining to processes or equipment not under the control of the central processor.

On-line
Pertaining to process equipment either under the control of the central processor or in direct communication with it.

Operand
Data used in a machine operation.

Operating System
A master program that controls all the resources of the computer and allocates them in an efficient manner to the other programs which are being run. It is sometimes called the 'Executive' program.

P

Peripheral
Any device which is connected to the central processor. Examples are input and output devices and backing store.

PL/1
Programming Lanugage/1

PMOS
P channel Metal Oxide Semiconductor (uses positive charges to transmit data).

Program
A series of instructions which the computer follows when carrying out a task.

PROM
Programmable Read Only Memory.

R

Random Access Memory (RAM)
: Such memory has read and write capability. Any location can be accessed as fast as any other.

Read Only Memory (ROM)
: There is no facility to write to, or amend the data or program held in the memory.

Real-Time
: The rapid processing of data for immediate use in time to influence events at the input source.

Register
: A special store in the computer which is used for the rapid transfer of data.

S

Software
: Any program or group of programs.

SOS
: Silicon-on-Sapphire.

SRAM
: Static Random Access Memory.

SSI
: Small Scale Integration.

Stack
: Area of store used for temporary storage of data.

Store
: Devices within the computer into which data or programs in binary form can be stored, recalled and operated upon.

Syntax
: The grammar of a programming language.

Systems Analysis
: The investigation of and analysis of a system to prepare it for automatic processing.

T

Time Sharing
A system in which several operators, who can be remotely sited, use the computer in such a way that each has the impression that he has sole use of the computer.

Transducer
A device for converting signals from one form to another eg pressure, temperature etc to an electrical signal, or vice versa.

TTL
Transistor/Transistor Logic.

TTY
Teletype.

V

VHSIC
Very High Speed Integrated Circuit.

Visual Display Unit (VDU)
A peripheral unit in which data is displayed on a cathode ray tube or plasma panel. It is usually equipped with a keyboard and is used as a combined input/output device.

VLSI
Very Large Scale Integration.

Volatile
Data is lost when power is removed.

W

Word
The unit of information used in a computer. The number of bits in a word is dependent upon machine architecture.

Bibliography

F G Pagan	A Practical Guide to Algol 68 Wiley 1976
N Graham	Introduction to Pascal West Publishing Co 1980
D Lewin	Theory and Design of Digital Computer Systems Nelson 1980
D E Heffer G A King D Keith	Basic Principles and Practice of Microprocessors Arnold 1980
A J T Colin	Programming for Microprocessors Newnes-Butterworths 1979
C A Ogdin	Software Design for Microcomputers Prentice-Hall 1978
D K Hsiao D S Kerr S E Madnick	Computer Security Academic Press 1979
H Ledgard	Ada: an Introduction / Ada Reference Manual (July 1980) Springer-Verlag 1981
H J Podell M Weiss	Tutorial : Business and Computers IEEE 1981
L H Putnam	Tutorial–Software Cost Estimating and Life-Cycle Control : Getting the Software Numbers IEEE 1980
D J Reifer	Tutorial : Sofrware Management IEEE 1979
C V Ramamoorthy R T Yeh	Tutorial : Software Methodology IEEE 1978

F Bates
M L Douglas
Programming Language/One : With Structured Programming
Prentice-Hall 1975

J B Maginnis
Fundamental ANSI COBOL Programming
Prentice-Hall 1975

H M Deitel
Introduction to Computer Programming with the BASIC Language
Prentice-Hall 1977

Index